INTRODUCTION TO GROUP CHARACTERS

INTRODUCTION TO A RICH CHARACTERS

INTRODUCTION TO GROUP CHARACTERS

Second edition

WALTER LEDERMANN

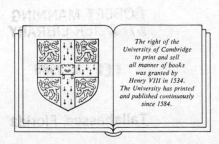

The right of the
University of Cambridge
to print and sell
all manner of books
was granted by
Henry VIII in 1534.
The University has printed
and published continuously
since 1584.

CAMBRIDGE UNIVERSITY PRESS

Cambridge

New York   New Rochelle   Melbourne   Sydney

Published by the Press Syndicate of the University of Cambridge
The Pitt Building, Trumpington Street, Cambridge CB2 1RP
32 East 57th Street, New York, NY 10022, USA
10 Stamford Road, Oakleigh, Melbourne 3166, Australia

First published 1977
Second edition 1987

Printed in Great Britain
at the University Press, Cambridge

British Library cataloguing in publication data
Ledermann, Walter
Introduction to group characters. – 2nd ed.
1. Finite groups 2. Characters of groups
I. Title
512′.22   QA171

Library of Congress cataloguing in publication data
Ledermann, Walter, 1911–
Introduction to group characters.

Bibliography
Includes index.
1. Characters of groups. I. Title.
QA171.L46 1987   512′.22   86-18766

ISBN 0 521 33246 X hard covers
ISBN 0 521 33781 X paperback

(First edition ISBN 0 521 21486 6 hard covers
ISBN 0 521 29170 4 paperback)

CONTENTS

PREFACE TO THE SECOND EDITION

I was pleased to accept the publisher's invitation to prepare an enlarged second edition of my book.

The theory of group representations is a large field, and it was difficult to decide what topics should fill the additional space that had become available.

Several colleagues were kind enough to give advice and also to draw my attention to errors in the first edition. I am particularly grateful to A. A. Bruen, R. W. Carter, W. L. Edge and G. D. James for their valuable comments.

However, it was impossible to incorporate all their suggestions. A selection had to be made: in accordance with the principal aim of this text, I have continued to place the emphasis on group characters rather than on the underlying representations.

The new material includes further work on tensor products, arithmetical properties of character values and the criterion for real representations due to Frobenius and Schur.

Once again, I should like to express my appreciation to the officers of Cambridge University Press for their help and understanding during the somewhat lengthy process of completing my task.

W. Ledermann

April, 1986

PREFACE

The aim of this book is to provide a straightforward introduction to the characters of a finite group over the complex field. The only prerequisites are a knowledge of the standard facts of Linear Algebra and a modest acquaintance with group theory, for which my text [13] would amply suffice. Thus the present volume could be used for a lecture course at the third-year undergraduate or at the post-graduate level.

The computational aspect is stressed throughout. The character tables of most of the easily accessible groups are either constructed in the text or are included among the exercises, for which answers and solutions are appended.

It goes without saying that a book on group characters must begin with an account of representation theory. This is now usually done in the setting of module theory in preference to the older approach by matrices. I feel that both methods have their merits, and I have formulated the main results in the language of either medium.

In this book I confine myself to the situation where representations are equivalent if and only if they have the same character. As soon as this fundamental fact is established, the emphasis shifts from the representations to the characters. Admittedly, some information is thereby sacrificed, and I had to be content with somewhat weaker versions of the theorems of A. H. Clifford [4] and G. W. Mackey [15]. However, character theory is sufficiently rich and rewarding by itself, and it leads to the celebrated applications concerning group structure without recourse to the underlying representations.

In the same vein, I have concentrated on the characters of the symmetric group rather than on its representations. The latter are expounded in the monographs of D. E. Rutherford [20] and G. de B. Robinson [19]. The cornerstone of our treatment is the generating function for the characters, due to Frobenius [9], whence it is easy to derive the Schur functions and their properties. On returning to Frobenius's original memoirs after many years I came to realise that familiarity with recondite results on determinants and symmetric functions that were common knowledge around 1900, could no longer be taken for granted in our

ix

time. I therefore decided to expand and interpret the classical master-pieces so as to make them self-contained without, I trust, spoiling the flavour of the creative power that permeated the early writings on this subject. In order to avoid unduly long digressions I relegated some of the auxiliary material to the Appendix.

There is a fairly extensive literature on representation theory, to which the reader may wish to turn for further instruction. Some of these books are listed in the Bibliography (p. 224). The substantial works of C. W. Curtis and I. Reiner [5] and L. Dornhoff [7] contain excellent bibliographies, which I do not wish to duplicate here. D. E. Littlewood's treatise [14] furnishes a great deal of valuable information, notably about the symmetric group.

My own interest in the subject goes back to an inspiring course by Issai Schur which I attended in 1931. This was subsequently published in the 'Zürich Notes' [21a]. Occasionally, Schur would enliven lectures with anecdotes about his illustrious teacher Frobenius, and I may be forgiven if I have succumbed to a bias in favour of an ancestral tradition.

My thanks are due to the Israel Institute of Technology (The Technion) at Haifa for permission to use a set of lecture notes prepared by their staff following a course I gave at their invitation in the spring of 1972. I am indebted to the University of Sussex for allowing me to include some examination questions among the exercises.

Finally, I wish to record my appreciation of the courtesy and patience which the Cambridge University Press has shown me during the preparation of this book.

W.L.

July, 1976

1

GROUP REPRESENTATIONS

1.1. Introduction

One of the origins of group theory stems from the observation that certain operations, such as permutations, linear transformations and maps of a space onto itself, permit of a law of composition that is analogous to multiplication. Thus the early work was concerned with what we may call concrete groups, in which the 'product' of two operations can be computed in every instance.

It was much later that group theory was developed from an axiomatic point of view, when it was realised that the structure of a group does not significantly depend upon the nature of its elements.

However, it is sometimes profitable to reverse the process of abstraction. This is done by considering homomorphisms

$$\theta: G \to \Gamma,$$

where G is an abstract group and Γ is one of the concrete groups mentioned above. Such a homomorphism is called a *representation of G*. Accordingly, we speak of representations by permutations, matrices, linear transformations and so on.

One of the oldest examples of a permutation representation is furnished by Cayley's Theorem, which states that a finite group

$$G: x_1(=1), x_2, \ldots, x_g$$

can be represented as a group of permutations of degree g, that is by permutations acting on g objects. In this case, the objects are the elements of G themselves. With a typical element x of G we associate the permutation

$$\pi(x) = \begin{pmatrix} x_1 & x_2 & \cdots & x_g \\ x_1 x & x_2 x & \cdots & x_g x \end{pmatrix}; \qquad (1.1)$$

this is indeed a permutation, because the second row in (1.1) consists of all the elements of G in some order. More briefly, we shall write

$$\pi(x): x_i \to x_i x \quad (i = 1, 2, \ldots, g).$$

1

If y is another element of G, we have analogously

$$\pi(y): x_i \to x_i y \quad (i = 1, 2, \ldots, g).$$

In this book the product of permutations is interpreted as a sequence of instructions read from left to right. Thus $\pi(x)\pi(y)$ signifies the operation whereby a typical element x_i of G is first multiplied by x and then by y on the right, that is

$$\pi(x)\pi(y): x_i \to (x_i x)y \quad (i = 1, 2, \ldots, g).$$

Since this is the same operation as

$$\pi(xy): x_i \to x_i(xy) \quad (i = 1, 2, \ldots, g),$$

we have established the crucial relationship

$$\pi(x)\pi(y) = \pi(xy), \tag{1.2}$$

which means that the map

$$\pi: G \to S_g$$

is a homomorphism of G into the symmetric group S_g, the group of all permutations on g symbols. This homomorphism, which is the content of Cayley's Theorem, is called the *right-regular representation* of G.

A given group G may have more than one representation by permutations, possibly of different degrees. Suppose that H is a subgroup of G of finite index n, and let

$$G = Ht_1 \cup Ht_2 \cup \ldots \cup Ht_n$$

be the coset decomposition of G relative to H. With a typical element x of G we associate the permutation

$$\sigma(x) = \begin{pmatrix} Ht_1 & Ht_2 & \ldots & Ht_n \\ Ht_1 x & Ht_2 x & \ldots & Ht_n x \end{pmatrix}, \tag{1.3}$$

in which the permuted objects are the n cosets. As before, it can be verified that

$$\sigma(x)\sigma(y) = \sigma(xy),$$

which proves that the map

$$\sigma: G \to S_n$$

is a homomorphism of G into S_n.

2

These examples serve to illustrate the notion of a permutation representation. For the remainder of the book we shall be concerned almost exclusively with homomorphisms

$$A: G \to GL_m(K), \tag{1.4}$$

where $GL_m(K)$ is the general linear group of degree m over K, that is the set of all non-singular $m \times m$ matrices with coefficients in a given *ground field K*. The integer m is called the *degree* (or *dimension*) of the representation A. We describe the situation formally as follows:

Definition 1.1. *Suppose that with each element x of the group G there is associated an m by m non-singular matrix*

$$A(x) = (a_{ij}(x)) \quad (i, j = 1, 2, \ldots, m),$$

with coefficients in the field K, in such a way that

$$A(x)A(y) = A(xy) \quad (x, y \in G). \tag{1.5}$$

Then $A(x)$ is called a matrix representation of G of degree (dimension) m over K.

A brief remark about nomenclature is called for: in Analysis we frequently speak of a 'function $f(x)$', when we should say 'a function (or map) f which assigns the value $f(x)$ to x'. We are here indulging in a similar abuse of language and refer to 'the representation $A(x)$' instead of using the more correct but clumsy phrase 'the homomorphism $A: G \to GL_m(K)$ which assigns to x the matrix $A(x)$'. When it is convenient, we abbreviate this to 'the representation A'.

Some consequences of (1.5) may be noted immediately. Let $x = y = 1$. Then we have that

$$\{A(1)\}^2 = A(1).$$

Since $A(1)$ is non-singular, it follows that

$$A(1) = I,$$

the unit matrix of dimension m. Next, put $y = x^{-1}$. Then

$$A(x)A(x^{-1}) = I,$$

so that

$$A(x^{-1}) = (A(x))^{-1}. \tag{1.6}$$

3

We emphasise that a representation A need not be injective ('one-to-one'), that is it may happen that $A(x) = A(y)$ while $x \neq y$. The kernel of A consists of those elements u of G for which $A(u) = I$. The kernel is always a normal, possibly the trivial, subgroup of G [13, p. 67]. The representation is injective or *faithful* if and only if the kernel reduces to the trivial group $\{1\}$. For the equation $A(x) = A(y)$ is equivalent to

$$A(x)(A(y))^{-1} = A(xy^{-1}) = I,$$

and for a faithful representation this implies that $xy^{-1} = 1$, that is $x = y$.

When $m = 1$, the representation is said to be *linear*. In this case we identify the matrix with its sole coefficient. Thus a linear representation is a function on G with values in K, say

$$\lambda: G \to K$$

such that

$$\lambda(x)\lambda(y) = \lambda(xy). \tag{1.7}$$

Every group possesses the *trivial* (formerly called *principal*) *representation* given by the constant function

$$\lambda(x) = 1 \quad (x \in G). \tag{1.8}$$

A non-trivial example of a linear representation is furnished by the *alternating character* of the symmetric group S_n (for each $n > 1$). This is defined by

$$\zeta(x) = \begin{cases} 1 & \text{if } x \text{ is even} \\ -1 & \text{if } x \text{ is odd.} \end{cases}$$

The equation

$$\zeta(x)\zeta(y) = \zeta(xy)$$

expresses a well-known fact about the parity of permutations [13, p. 134].

Let $A(x)$ be a representation of G and suppose that

$$B(x) = T^{-1}A(x)T, \tag{1.9}$$

where T is a fixed non-singular matrix with coefficients in K. It is readily verified that

$$B(x)B(y) = B(xy),$$

so that $B(x)$, too, is a representation of G. We say the representations $A(x)$ and $B(x)$ are *equivalent over K*, and we write

$$A(x) \sim B(x).$$

In the relationship (1.9) the exact form of T is usually irrelevant, but it is essential that its coefficients lie in K. As a rule, we do not distinguish between equivalent representations, that is we are only interested in equivalence classes of representations.

1.2. G-modules

The notion of equivalence becomes clearer if we adopt a more geometric approach. We recall the concept of a linear map

$$\alpha: V \to W$$

between two vector spaces over K. Under this map the image of a vector $\mathbf{v}$ of V will be denoted by $\mathbf{v}\alpha$, the operator α being written on the right. The map is *linear* if for all $\mathbf{u}, \mathbf{v} \in V$ and $h, k \in K$ we have that

$$(h\mathbf{u} + k\mathbf{v})\alpha = h(\mathbf{u}\alpha) + k(\mathbf{v}\alpha). \tag{1.10}$$

The zero map, simply denoted by 0, is defined by $\mathbf{v}0 = \mathbf{0}$ for all $\mathbf{v} \in V$.

The idea of a linear map does not involve the way in which the vector spaces may be referred to a particular basis. However, in order to compute the image of individual vectors, it is usually necessary to choose bases for V and W. In this book we shall be concerned only with finite-dimensional vector spaces.
Let

$$\dim V = m, \qquad \dim W = n,$$

and write

$$V = [\mathbf{p}_1, \mathbf{p}_2, \ldots, \mathbf{p}_m], \quad W = [\mathbf{q}_1, \mathbf{q}_2, \ldots, \mathbf{q}_n] \tag{1.11}$$

to express that $\mathbf{p}_1, \mathbf{p}_2, \ldots, \mathbf{p}_m$ and $\mathbf{q}_1, \mathbf{q}_2, \ldots, \mathbf{q}_n$ are bases of V and W respectively.

The image of $\mathbf{p}_i$ under α is some vector in W and therefore a linear combination of the basis vectors of W. Thus we have a system of equations

$$\mathbf{p}_i\alpha = \sum_{j=1}^{n} a_{ij}\mathbf{q}_j \quad (i = 1, 2, \ldots, m), \tag{1.12}$$

where $a_{ij} \in K$. This information enables us to write down the image of any $\mathbf{v} \in V$ by what is known as the *principle of linearity*; for if

$$\mathbf{v} = \sum_{i=1}^{m} h_i\mathbf{p}_i,$$

the linearity property (1.10) implies that

$$\mathbf{v}\alpha = \sum_{i=1}^{m} h_i \mathbf{p}_i \alpha = \sum_{i=1}^{m} \sum_{j=1}^{n} h_i a_{ij} \mathbf{q}_j.$$

Hence we may state that the $m \times n$ matrix

$$A = (a_{ij})$$

describes the linear map α relative to the bases (1.11).

If we had used different bases, say

$$V = [\mathbf{p}_1', \mathbf{p}_2', \dots, \mathbf{p}_m'], \quad W = [\mathbf{q}_1', \mathbf{q}_2', \dots, \mathbf{q}_n'], \tag{1.13}$$

the same linear map α would have been described by the matrix

$$B = (b_{\lambda\mu}),$$

whose coefficients appear in the equations

$$\mathbf{p}_\lambda' \alpha = \sum_{\mu=1}^{n} b_{\lambda\mu} \mathbf{q}_\mu' \quad (\lambda = 1, 2, \dots, m). \tag{1.14}$$

The change of bases is expressed algebraically by equations of the form

$$\mathbf{p}_i = \sum_{\lambda=1}^{m} t_{i\lambda} \mathbf{p}_\lambda' \quad (i = 1, 2, \dots, m)$$

$$\mathbf{q}_j = \sum_{\mu=1}^{n} s_{j\mu} \mathbf{q}_\mu' \quad (j = 1, 2, \dots, n) \tag{1.15}$$

where $T = (t_{i\lambda})$ and $S = (s_{j\mu})$ are non-singular (invertible) matrices of dimensions m and n respectively. Inverting the first set of equations we write

$$\mathbf{p}_\lambda' = \sum_{i=1}^{m} \tilde{t}_{\lambda i} \mathbf{p}_i \quad (\lambda = 1, 2, \dots, m),$$

where $T^{-1} = (\tilde{t}_{\lambda i})$. The relationship between the matrices A and B can now be obtained as follows (for the sake of brevity we suppress the ranges of the summation suffixes):

$$\mathbf{p}_\lambda' \alpha = \sum_{i} \tilde{t}_{\lambda i} \mathbf{p}_i \alpha = \sum_{i,j} \tilde{t}_{\lambda i} a_{ij} \mathbf{q}_j = \sum_{i,j,\mu} \tilde{t}_{\lambda i} a_{ij} s_{j\mu} q_\mu',$$

whence on comparing this result with (1.14) we have that

$$B = T^{-1} A S. \tag{1.16}$$

In the present context we are concerned with the situation in which $V = W$ and α is invertible. Such a linear map

$$\alpha : V \to V$$

is called an *automorphism* of V over K. The matrix which describes α relative to any basis is non-singular; and any two matrices A and B which express α relative to two different bases are connected by an equation of the form

$$B = T^{-1}AT. \tag{1.17}$$

The set of all automorphisms of V over K forms a group which we denote by

$$\mathscr{A}_K(V),$$

or simply by $\mathscr{A}(V)$, when the choice of the ground field can be taken for granted. If α_1 and α_2 are two elements of $\mathscr{A}(V)$, their product $\alpha_1\alpha_2$ is defined by operator composition, that is, if $\mathbf{v} \in V$, then

$$\mathbf{v}(\alpha_1\alpha_2) = (\mathbf{v}\alpha_1)\alpha_2.$$

We now consider representations of G by automorphisms of a vector space V. Thus we are interested in homomorphisms

$$G \to \mathscr{A}_K(V). \tag{1.18}$$

This means that with each element of x of G there is associated an automorphism

$$\alpha(x): V \to V$$

in such a way that

$$\alpha(x)\alpha(y) = \alpha(xy) \quad (x, y \in G). \tag{1.19}$$

We call (1.18) an *automorphism representation* of G, with the understanding that a suitable vector space V over K is involved.

In order to compute $\alpha(x)$ we refer V to a particular basis, say

$$V = [\mathbf{p}_1, \mathbf{p}_2, \ldots, \mathbf{p}_m]. \tag{1.20}$$

Applying (1.12) to the case in which $V = W$ we find that the action of $\alpha(x)$ is described by a matrix

$$A(x) = (a_{ij}(x))$$

over K, where

$$\mathbf{p}_i \alpha(x) = \sum_{j=1}^{m} a_{ij}(x)\mathbf{p}_j \quad (i = 1, 2, \ldots, m). \tag{1.21}$$

7

By virtue of (1.19) the matrix function $A(x)$ satisfies

$$A(x)A(y) = A(xy).$$

When the basis of V is changed, $\alpha(x)$ is described by a matrix of the form

$$B(x) = T^{-1}A(x)T,$$

where T is a non-singular matrix over K which is independent of x. Thus a representation $\alpha(x)$ gives rise to a class of equivalent matrix representations $A(x), B(x), \ldots$. Conversely, if we start with a matrix representation $A(x)$ we can associate with it an automorphism representation $\alpha(x)$ by starting with an arbitrary vector space (1.20) and defining the action of $\alpha(x)$ by means of (1.21).

Summing up, we can state that the classes of equivalent matrix representations are in one-to-one correspondence with automorphism representations of suitable vector spaces.

It is advantageous to push abstraction one stage further. In an automorphism representation each element x of G is associated with an automorphism $\alpha(x)$ of V. We shall now denote this automorphism simply by x; in other words, we put

$$\mathbf{v}x = \mathbf{v}\alpha(x), \tag{1.22}$$

and we say that G acts on V in accordance with (1.22). Formally, this defines a right-hand multiplication of a vector in V by an element of G. It is convenient to make the following

Definition 1.2. *Let G be a group. The vector space V over K is called a G-module, if a multiplication $\mathbf{v}x$ ($\mathbf{v} \in V$, $x \in G$) is defined, subject to the rules:*

  (i) $\mathbf{v}x \in V$;
  (ii) $(h\mathbf{v} + k\mathbf{w})x = h(\mathbf{v}x) + k(\mathbf{w}x)$, ($\mathbf{v}, \mathbf{w} \in V$; $h, k \in K$);
  (iii) $\mathbf{v}(xy) = (\mathbf{v}x)y$;
  (iv) $\mathbf{v}1 = \mathbf{v}$.

Let us verify that, in an abstract guise, this definition recaptures the notion of an automorphism representation. Indeed, (i) states that multiplication by x induces a map of V into itself; (ii) expresses that this map is linear; (iii) establishes the homomorphic property (1.19); finally, (iii) and (iv) imply that x and x^{-1} induce mutually inverse maps so that all these maps are invertible.

If V is a G-module, we say that V *affords* the automorphism representation defined in (1.22) or else the matrix representation $A(x)$ given by

(1.21), except that we now write $\mathbf{p}_i x$ instead of $\mathbf{p}_i \alpha(x)$, thus

$$\mathbf{p}_i x = \sum_{j=1}^{m} a_{ij}(x)\mathbf{p}_j \qquad (i = 1, 2, \ldots, m). \qquad (1.23)$$

Representation theory can be expressed either in terms of matrices, or else in the more abstract language of modules. The foregoing discussion shows that the two methods are essentially equivalent. The matrix approach lends itself more readily to computation, while the use of modules tends to render the theory more elegant. We shall endeavour to keep both points of view before the reader's mind.

1.3. Characters

Let $A(x) = (a_{ij}(x))$ be a matrix representation of G of degree m. We consider the characteristic polynomial of $A(x)$, namely

$$\det(\lambda I - A(x)) = \begin{vmatrix} \lambda - a_{11}(x) & -a_{12}(x) & \ldots & -a_{1m}(x) \\ -a_{21}(x) & \lambda - a_{22}(x) & \ldots & -a_{2m}(x) \\ \ldots & \ldots & & \ldots \\ -a_{m1}(x) & -a_{m2}(x) & \ldots & \lambda - a_{mm}(x) \end{vmatrix}.$$

This is a polynomial of degree m in λ, and inspection shows that the coefficient of $-\lambda^{m-1}$ is equal to

$$\phi(x) = a_{11}(x) + a_{22}(x) + \ldots + a_{mm}(x).$$

It is customary to call the right-hand side of this equation the *trace* of $A(x)$, abbreviated to $\operatorname{tr} A(x)$, so that

$$\phi(x) = \operatorname{tr} A(x). \qquad (1.24)$$

We regard $\phi(x)$ as a function on G with values in K, and we call it the *character* of $A(x)$. If

$$B(x) = T^{-1}A(x)T \qquad (1.25)$$

is a representation equivalent to $A(x)$, then

$$\det(\lambda I - B(x)) = \det(\lambda I - A(x)), \qquad (1.26)$$

because

$$\lambda I - B(x) = T^{-1}(\lambda I - A(x))T,$$

whence (1.26) follows by taking determinants of each side. In particular, on comparing coefficients of λ^{m-1} in (1.26) we find that

$$b_{11}(x) + b_{22}(x) + \ldots + b_{mm}(x) = a_{11}(x) + a_{22}(x) + \ldots + a_{mm}(x),$$

that is, equivalent representations have the same character. Put in a different way, we can state that $\phi(x)$ expresses a property of the equivalence class of matrix representations of which $A(x)$ is a member; or again, $\phi(x)$ is associated with an automorphism representation of a suitable G-module. It is this invariant feature which makes the character a meaningful concept for our purpose.

Suppose that x and $y = t^{-1}xt$ are conjugate elements of G. Then in any matrix representation $A(x)$ we have that

$$A(y) = \left(A(t)\right)^{-1} A(x) A(t).$$

On taking traces on both sides and identifying $A(t)$ with T in (1.25) we find that

$$\operatorname{tr} A(y) = \operatorname{tr} A(x),$$

that is, by (1.24), $\phi(x) = \phi(y)$. Thus, in every representation, the character is constant throughout each conjugacy class of G. Accordingly, we say that ϕ is a *class function* on G.

For reference, we collect our main results:

Proposition 1.1. *Let $A(x)$ be a matrix representation of G. Then the character*

$$\phi(x) = \operatorname{tr} A(x)$$

has the following properties:
   (i) *equivalent representations have the same character;*
   (ii) *if x and y are conjugate in G, then $\phi(x) = \phi(y)$.*

1.4. Reducibility

As often happens, we gain insight into a mathematical structure by studying 'subobjects'. This leads us to the distinction between reducible and irreducible representations.

Definition 1.3. *Let V be a G-module over K. We say that U is a submodule of V if*
   (i) *U is a vector space (over K) contained in V, and*
   (ii) *U is a G-module, that is $\mathbf{u}x \in U$ for all $\mathbf{u} \in U$ and $x \in G$.*
Every G-module V possesses the trivial submodules $U = V$ and $U = 0$. A non-trivial submodule is also called a *proper submodule*.

Definition 1.4. *A G-module is said to be reducible over K if it possesses a proper submodule; otherwise it is said to be irreducible over K.*

10

It must be emphasised that reducibility refers to a particular ground field. A change of the ground field may render an irreducible G-module reducible or *vice versa*.

Let us translate the notion of reducibility into matrix language. Thus suppose that the G-module V of dimension m possesses a submodule U of dimension r, where

$$0 < r < m.$$

Let $A(x)$ be the matrix representation (1.21) afforded by V relative to the original basis (1.20). We shall now refer V to a new basis, which is *adapted* to the subspace U. This process, which is known from Linear Algebra, consists in first choosing an arbitrary basis for U, say

$$U = [\mathbf{u}_1, \mathbf{u}_2, \ldots, \mathbf{u}_r], \tag{1.27}$$

and then extending this to a basis of V by suitable vectors $\mathbf{w}_1, \mathbf{w}_2, \ldots, \mathbf{w}_s$, where

$$r + s = m.$$

Thus V has a basis of the form

$$V = [\mathbf{u}_1, \mathbf{u}_2, \ldots, \mathbf{u}_r, \quad \mathbf{w}_1, \mathbf{w}_2, \ldots, \mathbf{w}_s]. \tag{1.28}$$

The relationship between the bases (1.20) and (1.28) is expressed by a matrix T as in (1.15) and, when the basis (1.28) is used, the action of x is described by the matrix $B(x)$ given in (1.25). However, we obtain more useful information about $B(x)$ if we study the action of x by operating directly on the vectors in (1.28) in analogy with (1.23). The important point is that U is itself a G-module. In particular, $\mathbf{u}_i x$ lies in U. Thus we have equations of the form

$$\mathbf{u}_i x = \sum_{j=1}^{r} c_{ij}(x)\mathbf{u}_j \quad (i = 1, 2, \ldots, r), \tag{1.29}$$

where $c_{ij}(x) \in K$. As regards $\mathbf{w}_h x$, we can only state that it is a vector in V and hence expressible in the form

$$\mathbf{w}_h x = \sum_{j=1}^{r} e_{hj}(x)\mathbf{u}_j + \sum_{k=1}^{s} d_{hk}(x)\mathbf{w}_k, \quad (h = 1, 2, \ldots, s) \tag{1.30}$$

where $e_{hj}(x), d_{hk}(x) \in K$. The coefficients of $B(x)$ can be gleaned from the right-hand sides of (1.29) and (1.30). Briefly, we infer that

$$B(x) = \begin{pmatrix} C(x) & 0 \\ E(x) & D(x) \end{pmatrix}, \tag{1.31}$$

11

where $C(x)$, $D(x)$ and $E(x)$ are matrices over K of dimensions $r \times \bar{r}$, $s \times s$ and $s \times r$ respectively. We can now formulate the matrix analogue of Definition 1.4.

Definition 1.5. *The matrix representation $A(x)$ is reducible over K if there exists a non-singular matrix T over K such that*

$$B(x) = T^{-1}A(x)T = \begin{pmatrix} C(x) & 0 \\ E(x) & D(x) \end{pmatrix}, \qquad (1.32)$$

for all $x \in G$.

Since $B(x)$ is a matrix representation, we have that, for all x and y in G,

$$\begin{pmatrix} C(xy) & 0 \\ E(xy) & D(xy) \end{pmatrix} = \begin{pmatrix} C(x) & 0 \\ E(x) & D(x) \end{pmatrix}\begin{pmatrix} C(y) & 0 \\ E(y) & D(y) \end{pmatrix}.$$

Expanding the matrix product on the right and comparing corresponding blocks on both sides we obtain that

$$C(xy) = C(x)C(y) \qquad (1.33)$$

$$D(xy) = D(x)D(y) \qquad (1.34)$$

$$E(xy) = E(x)C(y) + D(x)E(y). \qquad (1.35)$$

This shows that $C(x)$ and $D(x)$ are themselves matrix representations of G. The significance of (1.35) is obscure at the moment, but this relation will play a decisive part later on (p. 22).

Reverting to the module point of view we observe that $C(x)$ is the representation afforded by the G-module U referred to the basis (1.27). This is made plain by (1.29). But it is more difficult to construct a G-module for $D(x)$ which is related to U and V in a natural manner. To this end we consider the quotient space V/U. The reader is reminded that a vector space is, in particular, an additive Abelian group. Hence U is a (normal) subgroup of V, and the elements of V/U are all the cosets

$$\bar{\mathbf{v}} = U + \mathbf{v},$$

the composition law being

$$(U + \mathbf{v}_1) + (U + \mathbf{v}_2) = U + (\mathbf{v}_1 + \mathbf{v}_2). \qquad (1.36)$$

Moreover, V/U is a vector space over K by virtue of the definition

$$h(U + \mathbf{v}) = U + h\mathbf{v} \quad (h \in K). \qquad (1.37)$$

The 'natural' map

$$\pi: V \to V/U$$

assigns to each vector $\mathbf{v}$ the coset $U + \mathbf{v}$, in which it lies, that is

$$\mathbf{v}\pi = U + \mathbf{v}.$$

The equations (1.36) and (1.37) can now be rewritten as

$$\mathbf{v}_1\pi + \mathbf{v}_2\pi = (\mathbf{v}_1 + \mathbf{v}_2)\pi,$$

$$h(\mathbf{v}\pi) = (h\mathbf{v})\pi,$$

which imply that π is a homomorphism, that is

$$(h_1\mathbf{v}_1 + h_2\mathbf{v}_2)\pi = h_1(\mathbf{v}_1\pi) + h_2(\mathbf{v}_2\pi),$$

where $h_1, h_2 \in K$. The kernel of π is the space U, which is the zero element of V/U. Hence if π is applied to any relation between vectors of V, those vectors which lie in U are 'killed'. For example, suppose that

$$2\mathbf{u}_1 + 4\mathbf{v}_1 = 3\mathbf{u}_2 + \mathbf{v}_2,$$

where $\mathbf{u}_1, \mathbf{u}_2 \in U$. Then

$$4\mathbf{v}_1\pi = \mathbf{v}_2\pi.$$

Using (1.28) we can readily construct a basis for V/U; in fact, we assert that

$$V/U = [\mathbf{w}_1\pi, \mathbf{w}_2\pi, \ldots, \mathbf{w}_s\pi]. \tag{1.38}$$

For an arbitrary element of V/U is of the form $\mathbf{v}\pi$, where $\mathbf{v}$ is a suitable vector of V, say

$$\mathbf{v} = \sum_{i=1}^{r} a_i\mathbf{u}_i + \sum_{j=1}^{s} b_j\mathbf{w}_j$$

$(a_i, b_j \in K)$. On applying π we find that

$$\mathbf{v}\pi = \sum_{j=1}^{s} b_j(\mathbf{w}_j\pi).$$

Hence the vectors (1.38) certainly span V/U; moreover, they are linearly independent. For suppose that

$$\sum_{j=1}^{s} c_j(\mathbf{w}_j\pi) = \left(\sum_{j=1}^{s} c_j\mathbf{w}_j\right)\pi = 0.$$

Then $c_1\mathbf{w}_1 + c_2\mathbf{w}_2 + \ldots + c_s\mathbf{w}_s$ would lie in the kernel of π, that is in U, say

$$c_1\mathbf{w}_1 + c_2\mathbf{w}_2 + \ldots + c_s\mathbf{w}_s = d_1\mathbf{u}_1 + d_2\mathbf{u}_2 + \ldots + d_r\mathbf{u}_r.$$

But such a relation contradicts (1.28), unless all coefficients are zero. This proves (1.38).

13

The vector space V/U has one further property, which is of particular importance in our context: we may regard V/U as a G-module if we define the action of G by

$$(U+\mathbf{v})x = U + \mathbf{v}x. \tag{1.39}$$

First of all, we must check that this rule is 'well-defined': the left-hand side of (1.39) remains unchanged if we replace $\mathbf{v}$ by $\mathbf{u}+\mathbf{v}$, where $\mathbf{u}$ is an arbitrary vector of U; hence the same ought to be true for the right-hand side. Indeed,

$$U + (\mathbf{u}+\mathbf{v})x = U + \mathbf{u}x + \mathbf{v}x = U + \mathbf{v}x,$$

because $\mathbf{u}x \in U$. By verifying the conditions of Definition 1.2 the reader will easily show that V/U is indeed a G-module. We can express (1.39) more briefly by

$$(\mathbf{v}\pi)x = (\mathbf{v}x)\pi, \tag{1.40}$$

which states that π commutes with the action of G. Finally, applying π to (1.30) and using (1.40) we find that

$$(\mathbf{w}_h\pi)x = \sum_{k=1}^{s} d_{hk}(x)(\mathbf{w}_k\pi),$$

which shows that the G-module V/U affords the matrix representation $D(x)$ relative to the basis (1.38). This answers the query we raised on p. 12.

It may happen that the submodule U is irreducible; if not, it possesses a proper submodule U', which, of course, is also a submodule of V. Continuing in this manner we arrive at an irreducible submodule U_1 of V. Alternatively, we might have argued from the outset that, if V is reducible, it must possess a proper submodule of minimal dimension; such a minimal submodule is then necessarily irreducible. Thus, without loss of generality, we may assume that, if $A(x)$ is reducible, there exists a non-singular matrix T such that

$$T^{-1}A(x)T = \begin{pmatrix} A_1(x) & 0 \\ E(x) & D(x) \end{pmatrix}, \tag{1.41}$$

where $A_1(x)$ is irreducible. If now $D(x)$ is reducible, the process can be continued until we reach the situation described in the following theorem:

Theorem 1.1. *Let $A(x)$ be a matrix representation of G of degree m over K. Then either $A(x)$ is irreducible or else*

$$A(x) \sim \begin{pmatrix} A_1(x) & 0 & 0 & \dots & 0 \\ A_{21}(x) & A_2(x) & 0 & \dots & 0 \\ A_{31}(x) & A_{32}(x) & A_3(x) & \dots & 0 \\ \dots & \dots & \dots & & \dots \\ A_{l1}(x) & A_{l2}(x) & A_{l3}(x) & \dots & A_l(x) \end{pmatrix}, \qquad (1.42)$$

where $A_1(x), A_2(x), \dots, A_l(x)$ are irreducible over K.

Proof. We use induction with respect to m. When $m = 1$, $A(x)$ is certainly irreducible. Suppose now that $m > 1$ and that $A(x)$ is reducible. Hence we may assume that (1.41) holds. If $D(x)$ is irreducible, we have established (1.42) with $l = 2$. Otherwise, let

$$\dim A_1(x) = r, \qquad \dim D(x) = s.$$

By induction we assume that there exists a non-singular matrix S_0 over K of degree s such that

$$S_0^{-1} D(x) S_0 = \begin{pmatrix} A_2(x) & 0 & \dots & 0 \\ A_{32}(x) & A_3(x) & \dots & 0 \\ A_{l2}(x) & A_{l3}(x) & \dots & A_l(x) \end{pmatrix}, \qquad (1.43)$$

where $A_2(x), A_3(x), \dots, A_l(x)$ are irreducible. On putting

$$S = \begin{pmatrix} I & 0 \\ 0 & S_0 \end{pmatrix},$$

where I is the unit matrix of degree r, we find that

$$S^{-1} T^{-1} A(x) T S = \begin{pmatrix} A_1(x) & 0 \\ S_0^{-1} E(x) & S_0^{-1} D(x) S_0 \end{pmatrix}.$$

The proof is concluded by substituting (1.43) in this matrix and partitioning $S_0^{-1} E(x)$ as

$$\begin{pmatrix} A_{21}(x) \\ A_{31}(x) \\ \vdots \\ A_{l1}(x) \end{pmatrix}$$

to conform with the grouping of rows in (1.43).

Briefly, this theorem states that every matrix representation can be brought into lower triangular block form, in which the diagonal blocks are irreducible.

15

The module version of the same result runs as follows: if V is a reducible G-module, there exists a chain of submodules

$$0 = U_0 \subset U_1 \subset U_2 \subset \ldots \subset U_l = V \tag{1.44}$$

such that U_k/U_{k-1} ($k = 1, 2, \ldots, l$) is irreducible. The submodules are said to form a *composition series* for V. The existence of such a series is easy to establish. For if V is reducible we can choose a minimal proper submodule and start with the two-link chain

$$U_1 \subset V.$$

In analogy with a well-known result on quotient groups [11, p. 71], those submodules which lie between U_1 and V are in one-to-one correspondence with the submodules of V/U_1. Hence, unless V/U_1 is irreducible, there exists a submodule U_2 such that

$$U_1 \subset U_2 \subset V.$$

This process of refinement is continued until a composition series is obtained. We then choose a new basis for V which is adapted to the submodules U_1, U_2, U_3, ... More precisely, we choose a basis for U_1, then augment it so as to obtain a basis for U_2, extend it to a basis for U_3, and so on. When the new basis for V has been completed in this way, the matrix representation will appear in the form (1.42). This is, in fact, the situation covered by the *Jordan-Hölder Theorem* for 'groups with operators'. The most important result is the uniqueness up to isomorphism of the quotient modules U_k/U_{k-1}, apart from their order. In the matrix formulation this means that the diagonal blocks in (1.42) are uniquely determined by $A(x)$ and K, except for equivalence and rearrangement. We do not wish to lean on this theorem, which we have here quoted without proof, because we shall develop methods that are more appropriate for the cases we are going to consider in the sequel.

The rôle of the non-diagonal blocks in (1.42) is less evident, but no significance is attached to the fact that the zero blocks occur above rather than below the diagonal. This is related to our convention of writing operators on the right of the operand. If $A(x)$ is a matrix representation of G, so is the *contragredient representation*

$$A^{\dagger}(x) = A'(x^{-1}),$$

where the prime denotes transposition; indeed

$$A^{\dagger}(x)A^{\dagger}(y) = A'(x^{-1})A'(y^{-1})$$
$$= (A(y^{-1})A(x^{-1}))' = (A((xy)^{-1}))' = A^{\dagger}(xy).$$

Clearly,

$$A^{\dagger\dagger}(x) = A(x). \tag{1.45}$$

If $A(x)$ has been reduced to a lower triangular form, then $A^{\dagger}(x)$ appears in upper triangular form and *vice versa*. But (1.45) implies that every representation may be regarded as the contragredient of some representation.

1.5. Permutation representations

In this section we present a few elementary facts about permutation groups. A more detailed account will be given in Chapter 4. Let G be a permutation group of degree m, that is a subgroup of S_m, possibly the whole of S_m, and let

$$x = \begin{pmatrix} 1 & 2 & \dots & m \\ \xi_1 & \xi_2 & \dots & \xi_m \end{pmatrix} \tag{1.46}$$

be a typical element of G.

We construct a G-module V as follows: starting from an arbitrary vector space

$$V = [\mathbf{v}_1, \mathbf{v}_2, \dots, \mathbf{v}_m]$$

of dimension m over K, we define the action of G on V by the rule

$$\mathbf{v}_i x = \mathbf{v}_{\xi_i} \quad (i = 1, 2, \dots, m). \tag{1.47}$$

For elementary properties of permutations we refer the reader to texts on group theory, for instance [**13**, Chapter VII]. In particular, the symbol for x can be rewritten by listing the objects in the top row in any order whatever, provided that the total information is the same as that conveyed in (1.46), thus for example

$$\begin{pmatrix} 1 & 2 & 3 & 4 \\ 2 & 1 & 4 & 3 \end{pmatrix} = \begin{pmatrix} 4 & 2 & 1 & 3 \\ 3 & 1 & 2 & 4 \end{pmatrix} = \begin{pmatrix} 3 & 1 & 4 & 2 \\ 4 & 2 & 3 & 1 \end{pmatrix}.$$

If

$$y = \begin{pmatrix} 1 & 2 & \dots & m \\ \eta_1 & \eta_2 & \dots & \eta_m \end{pmatrix} = \begin{pmatrix} \xi_1 & \xi_2 & \dots & \xi_m \\ \tau_1 & \tau_2 & \dots & \tau_m \end{pmatrix}$$

is another element of G, then

$$(\mathbf{v}_i x)y = \mathbf{v}_{\xi_i} y = \mathbf{v}_{\tau_i}.$$

17

But

$$xy = \begin{pmatrix} 1 & 2 & \dots & m \\ \tau_1 & \tau_2 & \dots & \tau_m \end{pmatrix},$$

whence $v_i(xy) = v_{\tau_i}$. Hence $(v_i x)y = v_i(xy)$ for $i = 1, 2, \dots, m$. By linearity (p. 5), we have that $(vx)y = v(xy)$ for all $v \in V$. This is item (iii) in Definition 1.2. Since the other conditions are evidently satisfied, we conclude that V is indeed a G-module by virtue of (1.47). In the present case (1.23) reduces to (1.47). Hence V affords the representation

$$N(x) = (\delta_{\xi_i j}).$$

Each $N(x)$ is a permutation matrix, that is it has exactly one unit in each row and in each column, all other coefficients being zero. The representation is called the *natural representation* of G. Thus every permutation group possesses a natural representation.

It is easy to find the *natural character* $v(x)$ of G, that is the character of $N(x)$. We have that

$$v(x) = \operatorname{tr} N(x) = \sum_{i=1}^{m} \delta_{\xi_i i}.$$

Now $\delta_{\xi_i i}$ is zero unless $i = \xi_i$, in which case it is equal to unity. But $i = \xi_i$ means that the object i remains fixed by x. Hence

$$v(x) = number\ of\ objects\ fixed\ by\ x. \tag{1.48}$$

As an illustration let us determine the natural representation of the symmetric group S_3. In the usual cycle notation [13, p. 130] the six elements of S_3 are

$$1, \quad a = (12), \quad b = (13), \quad c = (23), \quad e = (123), \quad f = (132).$$

The action of S_3 on the vector space

$$V = [\mathbf{v}_1, \mathbf{v}_2, \mathbf{v}_3] \tag{1.49}$$

is explicitly given as follows:

$$(\mathbf{v}_1, \mathbf{v}_2, \mathbf{v}_3)1 = (\mathbf{v}_1, \mathbf{v}_2, \mathbf{v}_3)$$

$$(\mathbf{v}_1, \mathbf{v}_2, \mathbf{v}_3)a = (\mathbf{v}_2, \mathbf{v}_1, \mathbf{v}_3)$$

$$(\mathbf{v}_1, \mathbf{v}_2, \mathbf{v}_3)b = (\mathbf{v}_3, \mathbf{v}_2, \mathbf{v}_1)$$

$$(\mathbf{v}_1, \mathbf{v}_2, \mathbf{v}_3)c = (\mathbf{v}_1, \mathbf{v}_3, \mathbf{v}_2)$$

$$(\mathbf{v}_1, \mathbf{v}_2, \mathbf{v}_3)e = (\mathbf{v}_2, \mathbf{v}_3, \mathbf{v}_1)$$

$$(\mathbf{v}_1, \mathbf{v}_2, \mathbf{v}_3)f = (\mathbf{v}_3, \mathbf{v}_1, \mathbf{v}_2).$$

In order to obtain the corresponding matrices $N(x)$ we have to express the right-hand side of each of these equations linearly in terms of the basis elements (1.49) in their original order, and to note the coefficients that arise. Thus

$$N(1) = I, \quad N(a) = \begin{pmatrix} 0 & 1 & 0 \\ 1 & 0 & 0 \\ 0 & 0 & 1 \end{pmatrix}, \quad N(b) = \begin{pmatrix} 0 & 0 & 1 \\ 0 & 1 & 0 \\ 1 & 0 & 0 \end{pmatrix},$$

$$N(c) = \begin{pmatrix} 1 & 0 & 0 \\ 0 & 0 & 1 \\ 0 & 1 & 0 \end{pmatrix}, \quad N(e) = \begin{pmatrix} 0 & 1 & 0 \\ 0 & 0 & 1 \\ 1 & 0 & 0 \end{pmatrix}, \quad N(f) = \begin{pmatrix} 0 & 0 & 1 \\ 1 & 0 & 0 \\ 0 & 1 & 0 \end{pmatrix}.$$

The values of $\nu(x)$ are now read off, namely

$$\nu(1) = 3, \quad \nu(a) = \nu(b) = \nu(c) = 1, \quad \nu(e) = \nu(f) = 0.$$

This confirms (1.48). Also, we observe that a, b, c have the same character values as have e and f, in accordance with the fact that they are two sets of conjugate elements (Proposition 1.1 (ii), p. 10).

When $m \geqslant 2$, the natural representation is always reducible. In fact, since each x merely permutes the basis vectors of V, the vector

$$\mathbf{u} = \mathbf{v}_1 + \mathbf{v}_2 + \ldots + \mathbf{v}_m$$

has the property that

$$\mathbf{u}x = \mathbf{u} \quad (x \in G). \tag{1.50}$$

Hence the one-dimensional space

$$U = [\mathbf{u}]$$

is a submodule of V. Clearly, (1.50) implies that U affords the trivial representation. If we change the basis of V so as to include $\mathbf{u}$ as one of the basis vectors, we shall find, by (1.31), that

$$N(x) \sim \begin{pmatrix} 1 & 0 \\ E(x) & D(x) \end{pmatrix} = M(x) \tag{1.51}$$

say, where $D(x)$ is a representation of degree $m-1$ and $E(x)$ is an $(m-1) \times 1$ matrix.

Let us carry out this construction in the case of S_3. We choose the basis

$$V = [\mathbf{u}, \mathbf{v}_1, \mathbf{v}_2],$$

where

$$\mathbf{u} = \mathbf{v}_1 + \mathbf{v}_2 + \mathbf{v}_3.$$

19

The vector $\mathbf{v}_3$ will now have to be replaced by $\mathbf{u} - \mathbf{v}_1 - \mathbf{v}_2$. Relative to the new basis the action of G is as follows:

$$(\mathbf{u}, \mathbf{v}_1, \mathbf{v}_2)1 = (\mathbf{u}, \mathbf{v}_1, \mathbf{v}_2)$$

$$(\mathbf{u}, \mathbf{v}_1, \mathbf{v}_2)a = (\mathbf{u}, \mathbf{v}_2, \mathbf{v}_1)$$

$$(\mathbf{u}, \mathbf{v}_1, \mathbf{v}_2)b = (\mathbf{u}, \mathbf{u} - \mathbf{v}_1 - \mathbf{v}_2, \mathbf{v}_2)$$

$$(\mathbf{u}, \mathbf{v}_1, \mathbf{v}_2)c = (\mathbf{u}, \mathbf{v}_1, \mathbf{u} - \mathbf{v}_1 - \mathbf{v}_2)$$

$$(\mathbf{u}, \mathbf{v}_1, \mathbf{v}_2)e = (\mathbf{u}, \mathbf{v}_2, \mathbf{u} - \mathbf{v}_1 - \mathbf{v}_2)$$

$$(\mathbf{u}, \mathbf{v}_1, \mathbf{v}_2)f = (\mathbf{u}, \mathbf{u} - \mathbf{v}_1 - \mathbf{v}_2, \mathbf{v}_1).$$

Expressing this information in matrix language, in accordance with (1.23), we find the representation $M(x)$, which is equivalent to $N(x)$; for example

$$M(b) = \begin{pmatrix} 1 & 0 & 0 \\ 1 & -1 & -1 \\ 0 & 0 & 1 \end{pmatrix}, \qquad M(e) = \begin{pmatrix} 1 & 0 & 0 \\ 0 & 0 & 1 \\ 1 & -1 & -1 \end{pmatrix},$$

and similarly for all the other $M(x)$, each of which is of the form (1.51). For reference, we append a complete list of the matrices $D(x)$, which furnish a two-dimensional representation of S_3:

$$D(1) = I, \qquad D(a) = \begin{pmatrix} 0 & 1 \\ 1 & 0 \end{pmatrix}, \qquad D(b) = \begin{pmatrix} -1 & -1 \\ 0 & 1 \end{pmatrix},$$

$$D(c) = \begin{pmatrix} 1 & 0 \\ -1 & -1 \end{pmatrix}, \qquad D(e) = \begin{pmatrix} 0 & 1 \\ -1 & -1 \end{pmatrix}, \qquad D(f) = \begin{pmatrix} -1 & -1 \\ 1 & 0 \end{pmatrix}.$$

1.6. Complete reducibility

A reducible matrix representation $A(x)$ can be brought into the triangular shape (1.31), but it would clearly be more satisfactory if we could remove the off-diagonal block $E(x)$ by a further transformation. If this could be done, then we should have that

$$A(x) \sim \begin{pmatrix} C(x) & 0 \\ 0 & D(x) \end{pmatrix},$$

or written more concisely

$$A(x) \sim \mathrm{diag}\big(C(x), D(x)\big). \tag{1.52}$$

Unfortunately, this aim cannot always be achieved, as is shown by the following example: let

$$G: \ldots x^{-2}, x^{-1}, x^0 (= 1), x, x^2, \ldots$$

be the infinite cyclic group generated by x. Then

$$A(x^h) = \begin{pmatrix} 1 & 0 \\ h & 1 \end{pmatrix} \quad (h = 0, \pm 1, \pm 2, \ldots)$$

is a representation of G, because, as is easily verified,

$$A(x^h)A(x^k) = A(x^{h+k}).$$

But when $h \neq 0$, it is impossible to transform $A(x^h)$ into diagonal form; for both latent roots of $A(x^h)$ are equal to unity so that the diagonal form would have to be

$$\text{diag}(1, 1) = I,$$

which in turn would lead to the absurd conclusion that $A(x^h)$ equals I.

However, it is a remarkable fact that the diagonal form is attainable under very general conditions, which will be satisfied in all cases henceforth considered in this book.

Theorem 1.2. (*Maschke's Theorem*). *Let G be a finite group of order g, and let K be a field whose characteristic is either equal to zero or else is prime to g. Suppose that $A(x)$ is a matrix representation of G over K such that*

$$A(x) \sim \begin{pmatrix} C(x) & 0 \\ E(x) & D(x) \end{pmatrix}. \tag{1.53}$$

Then

$$A(x) \sim \text{diag}(C(x), D(x)).$$

Remark. The hypothesis about the characteristic of K ensures that K contains an element that can be identified with the rational number g^{-1}. This certainly holds when K is the complex field $\mathbb{C}$ or one of its subfields.

Proof. Since $A(x)$ may be replaced by an equivalent representation, there is no loss of generality in assuming that, in place of (1.53), we have that

$$A(x) = \begin{pmatrix} C(x) & 0 \\ E(x) & D(x) \end{pmatrix}, \tag{1.54}$$

where $C(x)$ and $D(x)$ are square matrices of degrees r and s respectively.

21

We seek a non-singular matrix T over K which is independent of x and has the property that

$$T^{-1}A(x)T = \begin{pmatrix} C(x) & 0 \\ 0 & D(x) \end{pmatrix}. \tag{1.55}$$

It turns out that we can solve the problem by putting

$$T = \begin{pmatrix} I_r & 0 \\ Q & I_s \end{pmatrix}, \qquad T^{-1} = \begin{pmatrix} I_r & 0 \\ -Q & I_s \end{pmatrix},$$

where Q is an $s \times r$ matrix that remains to be determined. Substituting for $A(x)$ from (1.54) and transposing the first factor in (1.55) we have that

$$\begin{pmatrix} C(x) & 0 \\ E(x) & D(x) \end{pmatrix}\begin{pmatrix} I_r & 0 \\ Q & I_s \end{pmatrix} = \begin{pmatrix} I_r & 0 \\ Q & I_s \end{pmatrix}\begin{pmatrix} C(x) & 0 \\ 0 & D(x) \end{pmatrix}.$$

On expanding the matrix products and comparing corresponding blocks on both sides we find that

$$E(x) + D(x)Q = QC(x) \tag{1.56}$$

is the sole condition for Q; of course, this has to hold simultaneously for all $x \in G$. We must now refer back to the equations (1.33), (1.34) and (1.35) which spell out that $A(x)$ is a representation of G. Multiplying (1.35) throughout by $C(y^{-1})$ we obtain that

$$E(xy)C(y^{-1}) = E(x) + D(x)E(y)C(y^{-1}).$$

We regard x as an arbitrary but fixed element of G, while y runs over G. Summing over y we find that

$$\sum_y E(xy)C(y^{-1}) = gE(x) + D(x)\sum_y E(y)C(y^{-1}). \tag{1.57}$$

On the left put $z = xy$ and note that when y runs over G so also does z. Hence

$$\sum_y E(xy)C(y^{-1}) = \sum_z E(z)C(z^{-1}x) = \left(\sum_z E(z)C(z^{-1})\right)C(x)$$

by virtue of (1.33). Thus, if we define

$$Q = \frac{1}{g}\sum_y E(y)C(y^{-1}),$$

the equation (1.57) becomes

$$QC(x) = E(x) + D(x)Q,$$

which is the same as (1.56). This completes the proof of Maschke's Theorem.

Let us formulate this result in the language of module theory. Suppose the G-module V affords the reducible representation $A(x)$. Then V possesses a proper submodule.

$$U = [\mathbf{u}_1, \mathbf{u}_2, \ldots, \mathbf{u}_r],$$

which affords the representation $C(x)$ in (1.53). In our previous discussion (p. 14) the interpretation of $D(x)$ required the use of V/U. But under the present condition the situation has been greatly simplified: Maschke's Theorem asserts that there exists a basis

$$V = [\mathbf{u}_1, \mathbf{u}_2, \ldots, \mathbf{u}_r, \mathbf{w}_1, \mathbf{w}_2, \ldots, \mathbf{w}_s], \tag{1.58}$$

relative to which all the coefficients $e_{hj}(x)$ in (1.30) are equal to zero so that

$$\mathbf{w}_h x = \sum_{k=1}^{s} d_{hk}(x)\mathbf{w}_k \quad (h = 1, 2, \ldots, s).$$

This means that

$$W = [\mathbf{w}_1, \mathbf{w}_2, \ldots, \mathbf{w}_s]$$

is a G-module which affords the representation $D(x)$. In these circumstances it is customary to write

$$V = U \oplus W$$

and to say that V is the *direct sum* of the G-modules U and W.

Maschke's Theorem can be applied at each stage of the process that leads to Theorem 1.1 (p. 15); thus all non-diagonal blocks in (1.42) can be reduced to zero by a suitable equivalence transformation. The result is usually expressed in terms of the following definition:

Definition 1.6. *A matrix representation $A(x)$ over K is said to be completely reducible if*

$$A(x) \sim \mathrm{diag}(A_1(x), A_2(x), \ldots, A_l(x)),$$

where $A_i(x)$ $(i = 1, 2, \ldots, l)$ are irreducible representations over K.

Again, the G-module V over K is said to be completely reducible, if

$$V = U_1 \oplus U_2 \oplus \ldots \oplus U_l,$$

where $U_i (i = 1, 2, \ldots, l)$ are irreducible G-modules over K.

Of course, the case $l = 1$ refers to an irreducible representation $A(x)$ (an irreducible G-module V).

Hence we have the more general version of Maschke's Theorem:

Theorem 1.3. *Let G be a finite group of order g, and let K be a field whose characteristic is zero or else prime to g. Then*
  (i) *every matrix representation of G over K is completely reducible, or equivalently,*
  (ii) *every G-module over K is completely reducible.*

In subsequent chapters of this book we shall be exclusively concerned with finite groups and with ground fields of characteristic zero. We shall therefore assume from now on that all representations are completely reducible.

1.7. Schur's Lemma

This celebrated result was enunciated by Issai Schur in his classical paper [21], which provided a more accessible approach to group characters than that given by Frobenius, who was the founder of the subject. The lemma is not particularly difficult to prove, but Schur recognised its importance for developing the theory. Indeed, it may be claimed that Maschke's Theorem and Schur's Lemma are two pillars on which the edifice of representation theory rests.

We begin by presenting the original matrix version used by Schur.

Theorem 1.4. *(Schur's Lemma). Let $A(x)$ and $B(x)$ be two irreducible representations over K of a group G, and suppose there exists a constant matrix T over K such that*

$$TA(x) = B(x)T \tag{1.59}$$

for all x in G. Then (i) either $T = 0$ or (ii) T is non-singular so that $A(x) = T^{-1}B(x)T$, that is $A(x)$ and $B(x)$ are equivalent.

Rather than following Schur's matrix treatment, we prefer to translate the assertion into the language of G-modules, as this will provide a better understanding of the situation. To this end, we require the notion of a *G-homomorphism*.

Definition 1.7. *Let V and U be G-modules over K, not necessarily irreducible. The linear map*

$$\theta: V \to U$$

is called a G-homomorphism if it commutes with all the linear maps

24

induced by the action of G, that is if

$$(vx)\theta = (v\theta)x \tag{1.60}$$

for all $v \in V$ and $x \in G$.

Suppose now that U and V afford the matrix representations $A(x)$ and $B(x)$ relative to the bases

$$U = [\mathbf{u}_1, \mathbf{u}_2, \ldots, \mathbf{u}_m] \quad \text{and} \quad V = [\mathbf{v}_1, \mathbf{v}_2, \ldots, \mathbf{v}_n]$$

respectively. Thus

$$\mathbf{u}_i x = \sum_{j=1}^{m} a_{ij}(x)\mathbf{u}_j \quad (i = 1, 2, \ldots, m)$$

and

$$\mathbf{v}_h x = \sum_{k=1}^{n} b_{hk}(x)\mathbf{v}_k \quad (h = 1, 2, \ldots, n).$$

The linear map θ is described by an $n \times m$ matrix

$$T = (t_{hi})$$

whose coefficients appear in the equations

$$\mathbf{v}_h \theta = \sum_{i=1}^{m} t_{hi}\mathbf{u}_i \quad (h = 1, 2, \ldots, n) \tag{1.61}$$

(see (1.12), p. 5). Examining (1.60) for a typical basis vector $\mathbf{v}_h$ we obtain that

$$(\mathbf{v}_h x)\theta = \left(\sum_k b_{hk}(x)\mathbf{v}_k\right)\theta = \sum_{j,k} b_{hk}(x)t_{kj}\mathbf{u}_j,$$

and

$$(\mathbf{v}_h \theta)x = \left(\sum_i t_{hi}\mathbf{u}_i\right)x = \sum_{i,j} t_{hi}a_{ij}(x)\mathbf{u}_j.$$

Hence the condition (1.60) for a G-homomorphism θ reduces to the matrix equation

$$TA(x) = B(x)T \quad (x \in G). \tag{1.62}$$

Conversely, if T satisfies this relation, we can define θ by (1.61) and, by linearity, recover (1.60). Thus (1.60) and (1.62) are essentially equivalent, and we denote the correspondence between θ and T briefly by

$$\theta \leftrightarrow T, \tag{1.63}$$

assuming that definite bases have been chosen for V and U.

We recall that if $\theta: V \rightarrow U$ is a linear map, then the *kernel* of θ, that is

$$\ker \theta = \{\mathbf{v}_0 \in V | \mathbf{v}_0 \theta = \mathbf{0}\},$$

is a subspace of V, and the *image* of θ, that is

$$\operatorname{im} \theta = \{\boldsymbol{u}_1 \in U | \mathbf{u}_1 = \mathbf{v}_1 \theta \quad \text{for some } \mathbf{v}_1\},$$

is a subspace of U. The map θ is called *injective* if $\ker \theta = \{\mathbf{0}\}$, and it is said to be *surjective* if $\operatorname{im} \theta = U$. If θ is both injective and surjective, it is termed *bijective* or an isomorphism; in that case, the matrix T associated with θ is non-singular.

These concepts apply, of course, to G-homomorphisms, since they are, in particular, linear maps. But we then have the important additional property that both $\ker \theta$ and $\operatorname{im} \theta$ are themselves G-modules. For suppose that $\mathbf{v}_0 \theta = \mathbf{0}$; then $(\mathbf{v}_0 x)\theta = (\mathbf{v}_0 \theta)x = \mathbf{0}$, which proves that $\ker \theta$ is a G-module. Similarly, if $\mathbf{u}_1 = \mathbf{v}_1 \theta$, it follows that $\mathbf{u}_1 x = (\mathbf{v}_1 \theta)x = (\mathbf{v}_1 x)\theta$, which shows that $\operatorname{im} \theta$ is a G-module.

We can now formulate the abstract version of Schur's Lemma, which is equivalent to Theorem 1.4 by virtue of (1.63).

Theorem 1.4′. (*Schur's Lemma*). *Suppose that U and V are irreducible G-modules over K. Then a G-homomorphism*

$$\theta: V \rightarrow U$$

is (i) *either the zero map, or else* (ii) *θ is an isomorphism.*

Proof. It suffices to prove the following statement: if $\theta \neq 0$, then θ is a bijection, that is $\ker \theta = \{\mathbf{0}\}$ and $\operatorname{im} \theta = U$. Indeed, since $\ker \theta$ is a submodule of the irreducible G-module V, we must have that either $\ker \theta = \{\mathbf{0}\}$ or $\ker \theta = V$. However, the second alternative would amount to saying that $\theta = 0$; hence we infer that $\ker \theta = \{\mathbf{0}\}$. Next, since $\operatorname{im} \theta$ is a submodule of U, either $\operatorname{im} \theta = \{\mathbf{0}\}$ or $\operatorname{im} \theta = U$; as the first possibility has been excluded, it follows that $\operatorname{im} \theta = U$. This proves Schur's Lemma.

If $A(x) = B(x)$, the conclusions of Schur's Lemma are particularly striking when the field K is algebraically closed, that is when every polynomial equation

$$a_0 x^r + a_1 x^{r-1} + \ldots + a_{r-1} x + a_r = 0$$

with coefficients in K has at least one root in K. The field $\mathbb{C}$ of complex numbers certainly has this property, and in the sequel we shall usually confine ourselves to this case.

26

Suppose now that $A(x)$ is irreducible over the algebraically closed field K, and let T be a matrix (necessarily square) over K which satisfies

$$TA(x) = A(x)T$$

for all $x \in G$. Then if k is any 'scalar' (element of K) we have that

$$(kI - T)A(x) = A(x)(kI - T) \quad (x \in G). \tag{1.64}$$

Since K is algebraically closed, there exists a scalar k_0 such that

$$\det(k_0 I - T) = 0.$$

The matrix $(k_0 I - T)$ is therefore singular. Applying Schur's Lemma to (1.64) with $k = k_0$, we are forced to conclude that

$$T = k_0 I.$$

Conversely, it is obvious that every matrix of the form kI $(k \in K)$ commutes with all the $A(x)$.

Corollary to Schur's Lemma. *Let $A(x)$ be an irreducible matrix representation of G over an algebraically closed field. Then the only matrices which commute with all the matrices $A(x)$ $(x \in G)$ are the scalar multiples of the unit matrix.*

1.8. The commutant (endomorphism) algebra

It would be interesting to know whether the converse of the corollary to Schur's Lemma is true, that is whether the fact that only the scalar multiples of the unit matrix commute with $A(x)$ implies that $A(x)$ is irreducible. This is part of a more general problem: suppose that $A(x)$ is a completely reducible matrix representation of G over the algebraically closed field K, and let $\mathscr{C}(A)$ be the set of all matrices T over K such that

$$TA(x) = A(x)T \tag{1.65}$$

for all $x \in G$. We note that, if T_1 and T_2 satisfy (1.65), so do the matrices $k_1 T_1 + k_2 T_2 (k_1, k_2 \in K)$ and $T_1 T_2$. Hence $\mathscr{C}(A)$ is an *algebra* over K, that is a vector space endowed with (associative) multiplication; in fact, it is a subalgebra of the complete matrix algebra $\mathscr{M}_m(K)$, where m is the degree of A. We call $\mathscr{C}(A)$ the *commutant algebra* of A. Let

$$B(x) = P^{-1}A(x)P$$

be a representation which is equivalent to $A(x)$. Then it is easy to see that $T \in \mathscr{C}(A)$ if and only if $P^{-1}TP \in \mathscr{C}(B)$. Moreover, the map

$$\psi \colon T \to P^{-1}TP$$

establishes an algebra-isomorphism between $\mathscr{C}(A)$ and $\mathscr{C}(B)$, because

$$(k_1 T_1 + k_2 T_2)\psi = k_1(T_1 \psi) + k_2(T_2 \psi),$$

$$(T_1 T_2)\psi = (T_1 \psi)(T_2 \psi).$$

Hence we have the following result.

Proposition 1.2. *If A and B are equivalent representations, then*

$$\mathscr{C}(A) \cong \mathscr{C}(B).$$

Since we shall be concerned only with properties that are common to isomorphic algebras, we may from the outset assume that

$$A(x) = \text{diag}(A_1(x), \ldots, A_s(x)), \qquad (1.66)$$

where the representations $A_1, A_2, \ldots, A_s$ are irreducible over K, though not necessarily inequivalent to one another. Let

$$m_i = \text{degree of } A_i \quad (i = 1, 2, \ldots, s).$$

In order to solve the equation (1.65) it is convenient to partition T into blocks, thus

$$T = \begin{pmatrix} T_{11} & T_{12} & \cdots & T_{1s} \\ T_{21} & T_{22} & \cdots & T_{2s} \\ \cdots & \cdots & & \cdots \\ T_{s1} & T_{s2} & \cdots & T_{ss} \end{pmatrix},$$

where T_{ij} is an m_i by m_j matrix. Substituting for $A(x)$ from (1.66) we find that (1.65) reduces to the s^2 equations

$$T_{ij}A_j(x) = A_i(x)T_{ij} \quad (i, j = 1, 2, \ldots, s)$$

for the matrix blocks. Schur's Lemma, together with the corollary, immediately provides the answer, namely

$$T_{ij} = \begin{cases} x_{ij} I_{m_i}, & \text{if } A_i \sim A_j \\ 0, & \text{if } A_i \not\sim A_j \end{cases}, \qquad (1.67)$$

where x_{ij} is an arbitrary element of K and I_{m_i} is the unit matrix of degree m_i. Conversely, any matrix of the form (1.67) satisfies (1.65). The matter is therefore settled in principle.

However, we wish to obtain a more precise description of $\mathscr{C}(A)$, taking account of the exact number of mutually inequivalent representations in (1.66) and of the multiplicities with which they occur. By a further equivalence transformation on A we may render equivalent representations equal and we can gather equal diagonal blocks together. Thus, without loss of generality, it may be assumed that

$$A(x) = \text{diag}(\underbrace{F_1(x), \ldots, F_1(x)}_{e_1}, \underbrace{F_2(x), \ldots, F_2(x)}_{e_2}, \ldots, \underbrace{F_l(x), \ldots, F_l(x)}_{e_l}),$$ (1.68)

where $F_1, F_2, \ldots, F_l$ are inequivalent irreducible representations of degrees $f_1, f_2, \ldots, f_l$, which occur with multiplicities $e_1, e_2, \ldots, e_l$ in the diagonal form of A.

At this point it is convenient to introduce the *Kronecker product* (*tensor* or *direct product*) of matrices as an aid to notation. We are here concerned only with square matrices: let $P = (p_{ij})$ and Q be matrices of degrees m and n respectively. Then we define the $mn \times mn$ matrix

$$P \otimes Q = \begin{pmatrix} p_{11}Q & p_{12}Q & \cdots & p_{1m}Q \\ p_{21}Q & p_{22}Q & \cdots & p_{2m}Q \\ \cdots & \cdots & & \cdots \\ p_{m1}Q & p_{m2}Q & \cdots & p_{mm}Q \end{pmatrix} = (p_{ij}Q).$$

Similarly, if

$$R \otimes S = (r_{ij}S),$$

where R is of degree m and S is of degree n, then the (i, j)th block in the product $(P \otimes Q)(R \otimes S)$ is

$$\left(\sum_{k=1}^{m} p_{ik} r_{kj} \right) QS.$$

Hence we have the important rule that

$$(P \otimes Q)(R \otimes S) = PR \otimes QS.$$ (1.69)

In particular, we have that

$$I \otimes Q = \text{diag}(Q, Q, \ldots, Q).$$

While we are on the subject of notation, we mention the *direct sum* of matrices. This is an alternative expression for a diagonal block matrix. Thus we shall say that the matrix

$$A = \text{diag}(A_1, A_2, \ldots, A_s)$$

is the direct sum of the matrices $A_1, A_2, \ldots, A_s$, and we shall write

$$A = A_1 \oplus A_2 \oplus \ldots \oplus A_s = \sum_{i=1}^{s} \oplus A_i.$$

This accords with the concept of the direct sum of G-modules, to which we referred in Definition 1.6. Indeed, the direct sum of G-modules affords a representation which is the direct sum of the representations afforded by its summands.

We continue our study of the commutant algebra, and begin with the special case of a representation

$$D(x) = \mathrm{diag}(F(x), \ldots, F(x)) = I_e \otimes F(x),$$

in which a single irreducible representation of degree f is repeated e times. By (1.67), the most general matrix T satisfying

$$TD(x) = D(x)T \quad (x \in G)$$

is of the form

$$T = \begin{pmatrix} x_{11}I_f & x_{12}I_f & \ldots & x_{1e}I_f \\ x_{21}I_f & x_{22}I_f & \ldots & x_{2e}I_f \\ \ldots & \ldots & & \ldots \\ x_{e1}I_f & x_{e2}I_f & \ldots & x_{ee}I_f \end{pmatrix} = X \otimes I_f,$$

where $X = (x_{ij})$ is an arbitrary $e \times e$ matrix over K. Hence we have that

$$\mathscr{C}(D) = \mathscr{M}_e \otimes I_f, \tag{1.70}$$

where $\mathscr{M}_e$ is the complete matrix algebra over K of degree e. If $T_1 = X_1 \otimes I_f$ and $T_2 = X_2 \otimes I_f$ are two elements of $\mathscr{C}(D)$, then, by the properties of the tensor product,

$$k_1 T_1 + k_2 T_2 = (k_1 X_1 + k_2 X_2) \otimes I_f,$$

and

$$T_1 T_2 = (X_1 X_2) \otimes I_f.$$

Hence, as an algebra, $\mathscr{C}(D)$ has the same structure as $\mathscr{M}_e$; the factor I_f in the tensor product merely 'blows up' the matrices, but does not affect the abstract laws of composition. Thus we can state that

$$\mathscr{C}(D) \cong \mathscr{M}_e. \tag{1.71}$$

It is now an easy matter to proceed to the general case of (1.65). We rewrite (1.68) as

$$A(x) = \sum_{i=1}^{l} \oplus (I_{e_i} \otimes F_i(x)).$$

If T is partitioned accordingly, all blocks of T which correspond to distinct summands of A are zero, while those which are associated with the same summand are given by (1.70). Hence

$$\mathscr{C}(A) = \sum_{i=1}^{l} \oplus (\mathcal{M}_{e_i} \otimes I_{f_i}). \tag{1.72}$$

The algebraic structure is described by

$$\mathscr{C}(A) \cong \sum_{i=1}^{l} \oplus \mathcal{M}_{e_i}.$$

Now $\mathcal{M}_e$ is a vector space over K of dimension e^2. Since in a direct sum of vector spaces the dimensions are added, we obtain that

$$\dim \mathscr{C}(A) = e_1^2 + e_2^2 + \ldots + e_l^2. \tag{1.73}$$

It still remains to find a relevant interpretation of the integer l, which, we recall, is the number of inequivalent irreducible constituents of A. As a preliminary, we ask the following simple question: what is the centre of $\mathcal{M}_e$, that is what is the set of matrices Z satisfying

$$ZX = XZ \tag{1.74}$$

for all $e \times e$ matrices X? The reader will probably know the answer; or, if not, he will certainly guess it: the centre of $\mathcal{M}_e$ consists of all scalar matrices. For the sake of completeness we indicate a proof. If $Z = (z_{ij})$ lies in the centre of $\mathcal{M}_e$, then, in particular, (1.74) holds when we take for X the matrix

$$W = \operatorname{diag}(w_1, w_2, \ldots, w_e),$$

where $w_1, w_2, \ldots, w_e$ are distinct elements of K. This implies that

$$z_{ij} w_j = w_i z_{ij} \quad (i, j = 1, 2, \ldots, e),$$

whence $z_{ij} = 0$ if $i \neq j$. Thus Z must be a diagonal matrix. Let

$$Z = \operatorname{diag}(z_1, z_2, \ldots, z_e).$$

Next we choose for X the permutation matrix

$$P = \begin{pmatrix} 0 & 1 & 0 & \ldots & 0 \\ 0 & 0 & 1 & \ldots & 0 \\ \ldots & & & & \ldots \\ 0 & 0 & 0 & \ldots & 1 \\ 1 & 0 & 0 & \ldots & 0 \end{pmatrix},$$

in which case (1.72) leads to

$$z_1 = z_2 = \ldots = z_e = c,$$

31

say, so that

$$Z = cI_f,$$

as asserted. Of course, every such matrix Z satisfies (1.74). Next, we determine the centre of $\mathscr{C}(A)$. A typical element of $\mathscr{C}(A)$ is a matrix of the form

$$T = \sum_{i=1}^{l} \oplus (X_i \otimes I_{f_i}),$$

where X_i is an arbitrary $e_i \times e_i$ matrix. Hence if

$$Z = \sum_{i=1}^{l} \oplus (C_i \otimes I_{f_i}) \tag{1.75}$$

lies in the centre of $\mathscr{C}(A)$, then

$$ZT = TZ$$

for all T in $\mathscr{C}(A)$. Using the rule (1.69) we see that this is equivalent to

$$C_i X_i = X_i C_i,$$

whence, by virtue of the foregoing discussion,

$$C_i = c_i I_{e_i} \quad (i = 1, 2, \ldots, l).$$

Substituting in (1.73) and noting that

$$c_i I_{e_i} \otimes I_{f_i} = c_i I_{e_i f_i},$$

we find that

$$Z = \sum_{i=1}^{l} \oplus c_i I_{e_i f_i}. \tag{1.76}$$

In every algebra the centre forms a subalgebra and therefore, in particular, a vector space. Now it is evident from (1.76) that the centre of $\mathscr{C}(A)$ is a vector space of dimension l; for $c_1, c_2, \ldots, c_l$ are arbitrary elements of K, and a basis for the centre of $\mathscr{C}(A)$ can be obtained by putting in turn one of the c's equal to unity and all other c's equal to zero. Thus

$$\dim \text{centre } \mathscr{C}(A) = l. \tag{1.77}$$

We summarise these results as follows:

Theorem 1.5. *Let $A(x)$ be a completely reducible representation of a group G over an algebraically closed field K, and suppose that*

$$A(x) = \sum_{i=1}^{l} \oplus (I_{e_i} \otimes F_i(x)), \tag{1.78}$$

where $F_1, F_2, \ldots, F_l$ are distinct inequivalent representations over K of degrees $f_1, f_2, \ldots, f_l$, occurring with multiplicities $e_1, e_2, \ldots, e_l$ in A. Let $\mathscr{C}(A)$ be the commutant algebra of A. Then

  (i) $\dim \mathscr{C}(A) = e_1^2 + e_2^2 + \ldots + e_l^2$,

  (ii) $\dim$ centre $\mathscr{C}(A) = l$,

  (iii) $m = e_1 f_1 + e_2 f_2 + \ldots + e_l f_l$,

where m is the degree of A.

Part (iii) is arrived at by counting the number of rows or columns on both sides of (1.78).

It goes without saying that the theory of the commutant algebra has its counterpart in the module language, the translation from one medium into the other being provided by the correspondence (1.63). Let V be a completely reducible G-module over V which affords the representation $A(x)$. Then a matrix T satisfying

$$TA(x) = A(x)T$$

corresponds to a *G-endomorphism* of V, that is a G-homomorphism

$$\theta: V \to V.$$

If θ_1 and θ_2 are G-endomorphisms, so also are $k_1 \theta_1 + k_2 \theta_2$ ($k_1, k_2 \in K$) and $\theta_1 \theta_2$, composition being defined in the usual way. Thus the set of all G-endomorphisms of V forms an algebra over K, which is often denoted by $\mathrm{Hom}_K(V, V)$. For the sake of brevity, we shall write

$$\mathscr{H}(V) = \mathrm{Hom}_K(V, V).$$

We call $\mathscr{H}(V)$ the *endomorphism algebra of V* over K. The equivalence of the two points of view is expressed by

$$\mathscr{H}(V) \cong \mathscr{C}(A),$$

and Theorem 1.5 can be recast as follows:

Theorem 1.5'. *Let V be a completely reducible G-module over K of dimension m, and suppose that*

$$V = U_1 \oplus \ldots \oplus U_1 \oplus U_2 \oplus \ldots \oplus U_2 \oplus \ldots \oplus U_l \oplus \ldots \oplus U_l,$$

where $U_1, U_2, \ldots, U_l$ are mutually non-isomorphic irreducible G-modules. Let $\dim U_i = f_i$ and let e_i be the multiplicity of U_i in V ($i = 1, 2, \ldots, l$). Then if $\mathscr{H}(V)$ is the endomorphism algebra of V, we have that

  (i) $\dim \mathscr{H}(V) = e_1^2 + e_2^2 + \ldots + e_l^2$,

  (ii) $\dim$ centre $\mathscr{H}(V) = l$,

  (iii) $m = e_1 f_1 + e_2 f_2 + \ldots + e_l f_l$.

In conclusion we remark that the algebra $\mathscr{C}(A)$ may be computed, at least in principle, by solving the linear equations (1.65) for the matrix T, although in practice the calculations are likely to be cumbersome. The information furnished in Theorem 1.5 will be of crucial importance in studying the representations of finite groups, to which we turn in the next chapter.

Exercises

1. Let H be a subgroup of G with coset representatives $t_1, t_2, \ldots, t_n$. Show that the kernel of the permutation representation (p. 2)

$$\sigma(x) = \begin{pmatrix} Ht_1 & Ht_2 & \ldots & Ht_n \\ Ht_1x & Ht_2x & \ldots & Ht_nx \end{pmatrix}$$

is the group

$$\bigcap_{i=1}^{n} t_i^{-1} H t_i.$$

2. Show that the map

$$a \to C(a) = \begin{pmatrix} 0 & 1 \\ -1 & -1 \end{pmatrix}$$

defines a representation of the cyclic group $\mathrm{gp}\{a \mid a^3 = 1\}$. Prove that this representation is irreducible over the field of real numbers.

3. Let ε be a linear map of the m-dimensional vector space V into itself such that

$$\varepsilon^2 = \varepsilon, \qquad \varepsilon \neq 0, \qquad \varepsilon \neq \iota \quad \text{(the identity map)}.$$

Define

$$U = \{\mathbf{u} \mid \mathbf{u}\varepsilon = \mathbf{u}\}, \qquad W = \{\mathbf{w} \mid \mathbf{w}\varepsilon = \mathbf{0}\}.$$

Show that U and W are subspaces of V which are invariant under ε, that is $U\varepsilon \subset U$, $W\varepsilon \subset W$.
Prove that

$$V = U \oplus W.$$

Deduce that if E is an $m \times m$ matrix such that

$$E^2 = E, \qquad E \neq 0, \qquad E \neq I,$$

there exists an integer r satisfying $1 \leqslant r < m$ and a non-singular matrix T such that

$$T^{-1}ET = \begin{pmatrix} I_r & 0 \\ 0 & 0 \end{pmatrix} = J.$$

4. Let G be a group. Suppose that with each $x \in G$ there is associated a matrix $A(x)$ in such a way that

$$A(x)A(y) = A(xy);$$

it is not assumed that $A(x)$ is non-singular. Prove that if $A(1)$ is singular but non-zero, then

$$A(x) \sim \begin{pmatrix} B(x) & 0 \\ 0 & 0 \end{pmatrix},$$

where $B(x)$ is a representation of G (consisting of non-singular matrices).

5. Let G be a permutation group of degree m (≥ 2). Suppose that $V = [\mathbf{v}_1, \mathbf{v}_2, \ldots, \mathbf{v}_m]$ affords the natural representation of G, that is $\mathbf{v}_i \sigma = \mathbf{v}_{i\sigma}$ ($\sigma \in G$; $i = 1, 2, \ldots, m$). Put $\mathbf{u}_j = \mathbf{v}_j - \mathbf{v}_m$ ($j = 1, 2, \ldots, m-1$). Show that

$$U = [\mathbf{u}_1, \mathbf{u}_2, \ldots, \mathbf{u}_{m-1}]$$

is an $(m-1)$-dimensional G-module.

In particular, when $m = 4$, compute the 3×3 matrices that describe the action of

$$\tau = (12), \quad \rho = (123), \quad \lambda = (12)(34), \quad \gamma = (1234)$$

on U and obtain the character value in each case.

6. Suppose the group G has a matrix representation $A(x)$ of degree two over the rationals, with the property that, for a certain central element z of G

$$A(z) = \begin{pmatrix} 0 & 1 \\ 1 & 0 \end{pmatrix}.$$

Prove that A is reducible over the rationals.

7. Let V be the m-dimensional vector space over $\mathbb{C}$, consisting of all row-vectors with m components. Suppose that U is an r-dimensional subspace of V ($r < m$), and define the orthogonal complement of U as

$$U^\perp = \{\mathbf{w} \in V \,|\, \mathbf{w}\bar{\mathbf{u}}' = \mathbf{0} \text{ for all } \mathbf{u} \in U\}.$$

Show that

$$V = U \oplus U^\perp.$$

Let $A(x)$ be a representation consisting of unitary matrices (briefly, a *unitary representation*) of a group, which need not be finite. Prove that $A(x)$ is completely reducible.

8. Let $A = (a_{ij})$ be an $m \times m$ matrix. Define

$$a^{(1^2)}(i, p; j, q) = a_{ij}a_{pq} - a_{iq}a_{pj},$$

where i, j, p, q range from 1 to m with the proviso that $i < p$ and $j < q$. Arrange the above expressions in a matrix $A^{(1^2)}$ of degree $\frac{1}{2}m(m-1)$, where the rows of $A^{(1^2)}$ are labelled by the pairs (i, p) and the columns by the pairs (j, q), each set of pairs being enumerated in lexical order, that is (i_1, p_1) precedes (i_2, p_2) if either $i_1 < i_2$, or $i_1 = i_2$ and $p_1 < p_2$. (This matrix is called the *second compound* of A, higher compounds being defined similarly.)

Prove that if B is another $m \times m$ matrix, then

$$(AB)^{(1^2)} = A^{(1^2)} B^{(1^2)}.$$

Deduce that if $\phi(x)$ is a character of a group G, so is the function

$$\tfrac{1}{2}\{(\phi(x))^2 - \phi(x^2)\} \quad (x \in G).$$

9. (*Converse of Schur's Lemma*) Let $A(x)$ be a representation of a finite group G over $\mathbb{C}$, having the property that only the scalar multiples of the unit matrix commute with all $A(x)$. Prove that $A(x)$ is irreducible.

[HINT: In theorem 1.5, dim $\mathscr{C}(A) = 1$. Hence one e is equal to unity and all the others are equal to zero.]

2

ELEMENTARY PROPERTIES OF GROUP CHARACTERS

2.1. Orthogonality relations

From now on we assume that G is a finite group of order g and that K is the field of complex numbers. Hence, by Maschke's Theorem, all representations of G are completely reducible. Furthermore, irreducibility henceforth means *absolute irreducibility*.

Let $A(x)$ and $B(x)$ be inequivalent irreducible representations of degrees f and f' respectively, and write

$$A(x) = (a_{ij}(x)) \quad (i, j = 1, 2, \ldots, f) \tag{2.1}$$

$$B(x) = (b_{pq}(x)) \quad (p, q = 1, 2, \ldots, f'). \tag{2.2}$$

For brevity, we adopt the convention in this section that i and j always run from 1 to f, and that p and q run from 1 to f'.

We choose $f'f$ arbitrary complex numbers ξ_{pj} and arrange them in an $f' \times f$ matrix

$$\Xi = (\xi_{pj}).$$

Next we construct the matrix

$$C = C(\Xi) = \sum_{y \in G} B(y^{-1}) \Xi A(y).$$

Let x be a fixed element of G; then $z = yx$ ranges over G when y does. Thus we may equally well write

$$C = \sum_{z} B(z^{-1}) \Xi A(z) = \sum_{y} B(x^{-1} y^{-1}) \Xi A(yx).$$

Since A and B are representations, we obtain that

$$C = B(x^{-1}) \left(\sum_{y} B(y^{-1}) \Xi A(y) \right) A(x),$$

or

$$B(x) C = C A(x) \quad (x \in G), \tag{2.3}$$

37

because $B(x^{-1}) = (B(x))^{-1}$. Applying Schur's Lemma to (2.3) we infer that C is the zero matrix, that is

$$\sum_{y \in G} \sum_{q,i} b_{pq}(y^{-1}) \xi_{qi} a_{ij}(y) = 0, \quad \text{for all } p \text{ and } j.$$

Since these equations hold for an arbitrary choice of Ξ, the coefficient of each ξ_{qi} must be zero, thus

$$\sum_{y \in G} b_{pq}(y^{-1}) a_{ij}(y) = 0, \quad \text{for all } i, j, p, q. \tag{2.4}$$

In order to express this and similar results more succinctly we introduce the notion of an *inner product* for functions on G with values in K. Let $\phi(x)$ and $\psi(x)$ be two such functions and write

$$\langle \phi, \psi \rangle = \frac{1}{g} \sum_{x \in G} \phi(x) \psi(x^{-1}). \tag{2.5}$$

Since this sum is unaltered if we replace x by x^{-1}, it follows that

$$\langle \phi, \psi \rangle = \langle \psi, \phi \rangle.$$

We shall say that the functions ϕ and ψ are *orthogonal* if

$$\langle \phi, \psi \rangle = 0.$$

The result (2.4) can now be expressed as

$$\langle a_{ij}, b_{pq} \rangle = 0, \quad \text{for all } i, j, p, q. \tag{2.6}$$

Thus if $A(x)$ and $B(x)$ are inequivalent irreducible representations, then every element of the matrix $A(x)$, when regarded as a function on G, is orthogonal to every element of $B(x)$.

Next, we consider a single irreducible representation of degree f. Then, as before, the matrix

$$C = \sum_y A(y^{-1}) \Xi A(y) \tag{2.7}$$

satisfies

$$A(x)C = CA(x) \quad (x \in G).$$

By the corollary to Schur's Lemma (p. 27)

$$C = \lambda I_f, \tag{2.8}$$

where $\lambda = \lambda\,(\Xi)$ is a scalar depending on Ξ. In order to evaluate λ we take the trace on both sides of (2.8), thus

$$\operatorname{tr} C = f\lambda. \tag{2.9}$$

Since $A(y^{-1}) = (A(y))^{-1}$, it follows that

$$\operatorname{tr}\left(A(y^{-1})\Xi A(y)\right) = \operatorname{tr}\Xi,$$

and all terms in the sum (2.7) have the same trace. Hence $\operatorname{tr} C = g\operatorname{tr}\Xi$, and

$$\lambda = \frac{g}{f}\operatorname{tr}\Xi.$$

Equating the expressions (2.7) and (2.8) for C we find that

$$\sum_y A(y^{-1})\Xi A(y) = \frac{g}{f}(\xi_{11} + \xi_{22} + \ldots + \xi_{ff})I_f,$$

whence on comparing coefficients of ξ_{jp}, we obtain that

$$\sum_y a_{ij}(y^{-1})a_{pq}(y) = 0, \quad \text{if } i \neq q \text{ or } j \neq p$$

and

$$\sum_y a_{ij}(y^{-1})a_{ji}(y) = \frac{g}{f}, \quad \text{for all } i \text{ and } j.$$

These results may be summarised by the statement that the coefficients of $A(x)$ satisfy the relations

$$\langle a_{ij}, a_{pq} \rangle = \frac{1}{f}\delta_{iq}\delta_{jp}. \tag{2.10}$$

The formulae (2.6) and (2.10) have important consequences for the characters of irreducible representations.

Theorem 2.1 (*Character relations of the first kind*). *Let $\chi(x)$ and $\chi'(x)$ be the characters of the irreducible representations $A(x)$ and $B(x)$ respectively. Then*

$$\langle \chi, \chi' \rangle = \begin{cases} 1 & \text{if } A(x) \sim B(x) \\ 0 & \text{if } A(x) \nsim B(x) \end{cases}.$$

Proof. Using the notations (2.1) and (2.2) we have that

$$\chi(x) = a_{11}(x) + a_{22}(x) + \ldots + a_{ff}(x),$$

$$\chi'(x) = b_{11}(x) + b_{22}(x) + \ldots + b_{f'f'}(x).$$

39

If A and B are inequivalent, we find that

$$\langle \chi, \chi' \rangle = \frac{1}{g} \sum_x \chi(x)\chi'(x^{-1}) = \sum_{i,p} \langle a_{ii}, b_{pp} \rangle = 0.$$

Next, if A and B are equivalent, we may suppose that $A = B$ because equivalent representations have the same character. In this case, by (2.10),

$$\langle \chi, \chi \rangle = \sum_{i,j} \langle a_{ii}, a_{jj} \rangle = \sum_{i,j} \frac{1}{f} \delta_{ij} \delta_{ij} = 1.$$

This proves the theorem.

The character of an irreducible representation is called a *simple character* while the character of a reducible representation is termed *compound*.

If G is a group of order g, then any group function ϕ on G may be regarded as a g-tuple, thus

$$\phi = (\phi(x_1), \phi(x_2) \ldots, \phi(x_g)),$$

where $x_1, x_2, \ldots, x_g$ are the elements of G enumerated in some fixed manner.

When viewed as g-tuples, any collection

$$\chi^{(1)}, \chi^{(2)}, \ldots, \chi^{(s)} \tag{2.11}$$

of distinct simple characters is linearly independent. For suppose that

$$\sum_{i=1}^{s} c_i \chi^{(i)} = 0,$$

where the right-hand side denotes the zero g-tuple. Taking the inner product of this equation with $\chi^{(j)}$, where $1 \leqslant j \leqslant s$, we find that

$$\sum_{i=1}^{s} c_i \langle \chi^{(i)}, \chi^{(j)} \rangle = \sum_{i=1}^{s} c_i \delta_{ij} = c_j = 0,$$

which proves the linear independence of (2.11). It is known from Linear Algebra that there can be no set comprising more than g linearly independent g-tuples. Therefore $s \leqslant g$, and it follows that a group of order g can have at most g inequivalent irreducible representations. For the time being we shall denote the precise number of irreducible representations by r. Let

$$\chi^{(1)}, \chi^{(2)}, \ldots, \chi^{(r)} \tag{2.12}$$

be a complete set of simple characters, and suppose that they correspond to the irreducible representations

$$F^{(1)}, F^{(2)}, \ldots, F^{(r)}, \qquad (2.13)$$

which, of course, are determined only up to equivalence. The degrees of the irreducible representations will be denoted by

$$f^{(1)}, f^{(2)}, \ldots, f^{(r)}$$

respectively.

By Maschke's Theorem, if A is an arbitrary representation, then

$$A \sim \mathrm{diag}(F, F', F'', \ldots),$$

where $F, F', F'', \ldots$ are irreducible. The constituents of A need not be distinct, but each of them can be chosen to be one of the complete set (2.13). If ϕ is the character of A, we can write

$$\phi = \sum_{j=1}^{r} d_j \chi^{(j)}, \qquad (2.14)$$

where $d_j (\geqslant 0)$ is the multiplicity of $F^{(j)}$ in A. On taking the inner product with $\chi^{(i)}$ we obtain that

$$d_i = \langle \phi, \chi^{(i)} \rangle \quad (i = 1, 2, \ldots, r). \qquad (2.15)$$

This is analogous to the way in which the coefficients of a Fourier series are determined. For this reason we refer to (2.14) as the *Fourier analysis* of ϕ or of A.

Let B be another representation, and suppose that its character, ψ, has the Fourier analysis

$$\psi = \sum_{j=1}^{r} e_j \chi^{(j)}, \qquad (2.16)$$

so that

$$e_i = \langle \psi, \chi^{(i)} \rangle \quad (i = 1, 2, \ldots, r). \qquad (2.17)$$

We are now in the position to establish the converse of the elementary result that equivalent representations have the same character. Indeed if

$$\phi(x) = \psi(x) \quad (x \in G),$$

then (2.15) and (2.17) immediately show that $d_i = e_i$ ($i = 1, 2, \ldots, r$). Hence A and B are equivalent to the same diagonal array of irreducible constituents and are therefore equivalent to each other. We record this important fact as follows:

41

Theorem 2.2. *Two representations of a finite group over the complex field are equivalent if and only if they have the same character.*

In other words, the trace of a representation furnishes us with complete information about the irreducible constituents into which this representation may be decomposed. In this sense, the character truly characterises a representation up to equivalence.

We append some further remarks about the Fourier analysis of a character. In the notation of (2.14) and (2.16) we have that

$$\langle \phi, \psi \rangle = \sum_{j=1}^{r} d_j e_j$$

and, in particular,

$$\langle \phi, \phi \rangle = \sum_{j=1}^{r} e_j^2. \tag{2.18}$$

Now a representation is irreducible if and only if it consists of a single, unrepeated, irreducible constituent, that is, we have

Proposition 2.1. *A representation with character ϕ is irreducible if and only if $\langle \phi, \phi \rangle = 1$.*

Example. The representation D of S_3, mentioned on p. 20 has the character ϕ, given by

$$\phi(1) = 2, \qquad \phi(a) = \phi(b) = \phi(c) = 0, \qquad \phi(e) = \phi(f) = -1.$$

Since

$$\langle \phi, \phi \rangle = \tfrac{1}{6}(4 + 0 + 0 + 0 + 1 + 1) = 1,$$

it follows that D is irreducible over the complex field.

2.2. The group algebra

We introduce an important new concept, which connects the structure of the group G with that of the underlying field K (in our case $K = \mathbb{C}$). As always,

$$G: x_1(=1), x_2, \ldots, x_g$$

is a finite multiplicative group, but we now assign to the elements of G a second rôle by regarding them as the basis vectors of a vector space

$$G_{\mathbb{C}} = [x_1, x_2, \ldots, x_g] \tag{2.19}$$

over the complex numbers. Thus a typical element of G_C is a 'formal' sum

$$v = \alpha_1 x_1 + \alpha_2 x_2 + \ldots + \alpha_g x_g$$

where $\alpha_1, \alpha_2, \ldots, \alpha_g$ are arbitrary complex numbers. Let

$$w = \beta_1 x_1 + \beta_2 x_2 + \ldots + \beta_g x_g$$

be another element of G_C. Then $v = w$ if and only if $\alpha_i = \beta_i$ $(i = 1, 2, \ldots, g)$, and we have the usual rules for a vector space, namely

$$v + w = \sum_{i=1}^{g} (\alpha_i + \beta_i) x_i,$$

$$\lambda v = \sum_{i=1}^{g} (\lambda \alpha_i) x_i \quad (\lambda \in \mathbb{C}), \qquad (2.20)$$

$$1v = v.$$

Furthermore, G_C possesses a multiplicative structure. For suppose that the multiplication table for G is expressed as

$$x_i x_j = x_{\mu(i,j)} \qquad (2.21)$$

where $\mu(i, j)$ is a well-defined integer lying between 1 and g. Multiplication in G_C is then defined by

$$vw = \sum_{i,j=1}^{g} \alpha_i \beta_j x_{\mu(i,j)}.$$

This multiplication is associative by virtue of the associative law for G. Thus, if $u, v, w \in G_C$, then

$$(uv)w = u(vw).$$

As we have already remarked (p. 27), a vector space which is endowed with an associative multiplication is called an algebra. Accordingly, we call G_C the *group algebra* of G over $\mathbb{C}$. The reader must bear with the slightly confusing situation caused by the fact that x may signify both an element of G and a vector of G_C. But there should be no logical difficulty, since G can be regarded as a finite subset of G_C consisting of those sums (2.20) in which one coefficient is unity while all the others are zero.

For the moment we exploit the multiplicative structure of G_C only in so far as it implies that

$$G_C x \subset G_C \quad (x \in G).$$

43

Thus $G_\mathbb{C}$ is a G-module of degree g over $\mathbb{C}$. Let $R(x)$ be the representation afforded by it. If we use the 'natural' basis (2.19), the action of x is described by

$$[x_1, x_2, \ldots, x_g]\, x = [x_1 x, x_2 x, \ldots, x_g x],$$

which is seen to be identical with the permutation representation $\pi(x)$ mentioned in (1.1). Employing the same term as on p. 2 we say that $G_\mathbb{C}$, as a G-module, affords the *right-regular representation* of G. Explicitly, if

$$R(x) = \bigl(r_{ij}(x)\bigr) \quad (i, j = 1, 2, \ldots, g),$$

then

$$x_i x = \sum_{j=1}^{g} r_{ij}(x) x_j.$$

On the other hand, if we put $x = x_s$ and use (2.21), we have that

$$x_i x_s = x_{\mu(i,s)}.$$

Hence

$$r_{ij}(x_s) = \delta_{\mu(i,s),j}. \tag{2.22}$$

Example. Find the right-regular representation of the four-group

$$V: 1, a, b, ab, \text{ where } a^2 = b^2 = 1, ab = ba.$$

Now

$$(1, a, b, ab)a = (a, 1, ab, b),$$

whence

$$R(a) = \begin{pmatrix} 0 & 1 & 0 & 0 \\ 1 & 0 & 0 & 0 \\ 0 & 0 & 0 & 1 \\ 0 & 0 & 1 & 0 \end{pmatrix}.$$

Similarly,

$$(1, a, b, ab)b = (b, ab, 1, a),$$

$$R(b) = \begin{pmatrix} 0 & 0 & 1 & 0 \\ 0 & 0 & 0 & 1 \\ 1 & 0 & 0 & 0 \\ 0 & 1 & 0 & 0 \end{pmatrix},$$

and

$$(1, a, b, ab)ab = (ab, b, a, 1),$$

$$R(ab) = \begin{pmatrix} 0 & 0 & 0 & 1 \\ 0 & 0 & 1 & 0 \\ 0 & 1 & 0 & 0 \\ 1 & 0 & 0 & 0 \end{pmatrix}.$$

It is easy to check that $R(a)R(b) = R(ab)$. Returning to the general case we denote the character of $R(x)$ by $\rho(x)$, thus

$$\rho(x) = \sum_{i=1}^{g} r_{ii}(x).$$

According to (1.48) we have that

$$\rho(x) = number\ of\ symbols\ fixed\ by\ x,$$

that is, for a particular x, $\rho(x)$ is the number of times that

$$x_i x = x_i \quad (i = 1, 2, \ldots, g)$$

holds. However, this equation is impossible unless $x = 1$. Hence $r_{ii}(x) = 0$ if $x \neq 1$ and $r_{ii}(1) = 1$ $(i = 1, 2, \ldots, g)$. Thus

$$\rho(1) = g, \qquad \rho(x) = 0 \quad \text{if } x \neq 1, \tag{2.23}$$

that is the character of R is given by the g-tuple

$$\rho = (g, 0, 0, \ldots, 0).$$

We now proceed to a detailed analysis of the right-regular representation. Suppose that

$$\rho = \sum_{i=1}^{r} e_i \chi^{(i)} \tag{2.24}$$

is the Fourier analysis of ρ. By (2.23)

$$e_i = \langle \rho, \chi^{(i)} \rangle = \frac{1}{g} \sum_{j=1}^{g} \rho(x_j) \chi^{(i)}(x_j^{-1}) = \chi^{(i)}(1),$$

that is

$$e_i = f^{(i)} \quad (i = 1, 2, \ldots, r).$$

Substituting in (2.24) we have that

$$\rho(x) = \sum_{i=1}^{r} f^{(i)} \chi^{(i)}(x), \tag{2.25}$$

which means that

$$R(x) \sim \sum_{i=1}^{r} \oplus \left(I_{f^{(i)}} \otimes F^{(i)}(x) \right).$$

This is the decomposition of $R(x)$ into irreducible constituents.

On putting $x = 1$ in (2.25) we obtain the interesting result that

$$g = \sum_{i=1}^{r} (f^{(i)})^2. \tag{2.26}$$

At this juncture one might put forward the view that the representation theory for finite groups, over the complex field, has been successfully accomplished. Indeed, the decomposition of $R(x)$ would bring to light all possible irreducible representations of the group, up to equivalence; in fact, each of these representations will appear with a multiplicity equal to its degree. However, this would be an impracticable suggestion, because the reduction of a sizeable representation is a very laborious operation. Moreover, we have still to discover the deeper relationships between representations and group structure. In particular, the significance of the number r needs to be explored.

Further information can be gleaned by applying Theorem 1.5′ (p. 33) to the G-module G_C. Thus we shall consider the algebra

$$\mathcal{H} = \mathrm{Hom}(G_C, G_C),$$

and, in particular, we shall compute dim $\mathcal{H}$ and dim centre $\mathcal{H}$. We recall that $\mathcal{H}$ consists of all those linear maps

$$\theta : G_C \to G_C$$

which commute with the action of G, that is

$$(v\theta)x = (vx)\theta \quad (v \in G_C, x \in G). \tag{2.27}$$

Let us first determine a necessary condition for such a map. Suppose the image of 1 under θ is the element t of G_C, that is

$$1\theta = t.$$

Substituting $v = 1$ in (2.27) and leaving x arbitrary we find that

$$tx = x\theta \quad (x \in G).$$

Hence, by applying this formula to each term of (2.20) we find that

$$v\theta = tv. \tag{2.28}$$

46

Conversely, if θ is defined by (2.28), where t is an arbitrary element of G_C, we may replace v by vx in (2.28) and obtain that

$$(vx)\theta = t(vx) = (tv)x = (v\theta)x,$$

so that (2.27) holds. Thus (2.28) sets up a one-to-one correspondence between the elements of $\mathcal{H}$ and those of G_C: every endomorphism of the G-module G_C is equivalent to the left-multiplication by a fixed element of G_C. Moreover, the correspondence (2.28) is an isomorphism between the vector spaces $\mathcal{H}$ and G_C. For, if analogously

$$v\eta = sv,$$

where $\eta \in \mathcal{H}$, $s \in G_C$, then

$$v(a\theta + b\eta) = (at + bs)v \quad (a, b \in \mathbb{C}).$$

It follows that

$$\dim \mathcal{H} = \dim G_C = g. \tag{2.29}$$

Now parts (i) and (iii) of Theorem 1.5' have been confirmed by (2.29) and (2.26).

A new result will emerge when we examine the centre of $\mathcal{H}$. Let ξ be a typical element of the centre of $\mathcal{H}$. Then

$$v\theta\xi = v\xi\theta \quad (v \in G_C, \theta \in \mathcal{H}). \tag{2.30}$$

We may suppose that θ is given by (2.28), where t ranges over G_C. Similarly, there exists a unique element z of G_C such that

$$v\xi = zv \quad (v \in G_C).$$

It is easily seen that (2.30) is equivalent to

$$ztv = tzv \quad (v \in G_C),$$

and since this holds for all v, we may put $v = 1$ and deduce that

$$zt = tz. \tag{2.31}$$

Let

$$t = \sum_{j=1}^{g} \alpha_j x_j,$$

where $\alpha_1, \alpha_2, \ldots, \alpha_g$ are arbitrary complex numbers. Hence (2.31) is equivalent to the conditions

$$zx_j = x_j z \quad (j = 1, 2, \ldots, g)$$

or

$$x_j^{-1} z x_j = z \quad (j = 1, 2, \ldots, g).$$

47

Since x_j can be any one of the elements of G, we can say that z lies in the centre of $G_\mathbb{C}$, and consequently ξ lies in the centre of $\mathcal{H}$, if and only if

$$u^{-1}zu = z \quad (u \in G). \tag{2.32}$$

Let

$$z = \sum_{i=1}^{g} \lambda_j x_j,$$

where $\lambda_1, \lambda_2, \ldots, \lambda_g$ are suitable complex numbers. Then (2.32) becomes

$$z = \sum_{j=1}^{g} \lambda_j x_j = \sum_{j=1}^{g} \lambda_j u^{-1} x_j u. \tag{2.33}$$

Thus conjugate basis elements have the same coefficient in z. Suppose the k conjugacy classes of G are listed as follows:

$$C_1 = \{1\}, \ C_2 = \{y_2, p_2^{-1} y_2 p_2, q_2^{-1} y_2 q_2, \ldots\}, \ldots,$$
$$C_\alpha = \{y_\alpha, p_\alpha^{-1} y_\alpha p_\alpha, q_\alpha^{-1} y_\alpha q_\alpha, \ldots\}, \ldots,$$
$$C_k = \{y_k, p_k^{-1} y_k p_k, q_k^{-1} y_k q_k, \ldots\}.$$

We associate with C_α the element

$$c_\alpha = y_\alpha + p_\alpha^{-1} y_\alpha p_\alpha + q_\alpha^{-1} y_\alpha q_\alpha + \ldots \tag{2.34}$$

of $G_\mathbb{C}$. On gathering the elements of C_α in (2.33) for each α and suitably renaming the coefficients we can write (2.33) as

$$z = \gamma_1 c_1 + \gamma_2 c_2 + \ldots + \gamma_k c_k, \tag{2.35}$$

where each γ is one of the λs. Conversely, we have that

$$u^{-1} c_\alpha u = c_\alpha \quad (\alpha = 1, 2, \ldots, k),$$

because the element on the left consists of the same terms as c_α, though possibly in a different order. Thus c_α lies in the centre of $G_\mathbb{C}$ and so therefore does (2.35), whatever the choice of the coefficients. Clearly, $c_1, c_2, \ldots, c_k$ are linearly independent because their terms are mutually disjoint. Hence

$$k = \dim \text{centre } G_\mathbb{C} = \dim \text{centre } \mathcal{H}.$$

Part (ii) of Theorem 1.5' now tells us that R involves precisely k inequivalent irreducible constituents. Since we already know that all possible irreducible representations, up to equivalence, occur in R, we have finally arrived at the conclusion that

$$r = k.$$

We summarise our results as follows:

Theorem 2.3. *Let G be a group of order g. If G has k conjugacy classes, there are, up to equivalence, k distinct irreducible representations over* $\mathbb{C}$,

$$F^{(1)}, F^{(2)}, \ldots, F^{(k)}.$$

If $F^{(i)}$ *is of degree* $f^{(i)}$, *then*

$$g = \sum_{i=1}^{k} (f^{(i)})^2.$$

2.3. The character table

The complete information about the characters of G is conveniently displayed in a character table, which lists the values of the k simple characters for all the elements. We know (p. 10) that a character is constant on each of the conjugacy classes. If $x \in C_\alpha$, we put $\chi(x) = \chi_\alpha$. Thus it is sufficient to record the values χ_α $(1 \le \alpha \le k)$. Denoting the number of elements in C_α by h_α, we have the class equation

$$h_1 + h_2 + \ldots + h_k = g. \tag{2.36}$$

Unless the contrary is stated, we adhere to the convention that $C_1 = \{1\}$ and that $F^{(1)}$ is the trivial representation

$$F^{(1)}(x) = 1 \quad (x \in G).$$

As in Theorem 2.3, the degree of $F^{(i)}$ will be denoted by $f^{(i)}$, so that

$$\chi_1^{(i)} = f^{(i)} \quad (i = 1, 2, \ldots, k).$$

Table 2.1 presents a typical character table. The body of the table is a $k \times k$ square matrix whose rows correspond to the different characters while each column contains the values of all simple characters for a particular conjugacy class.

G:

	C_1	C_2	$\ldots$	C_α	$\ldots$	C_k
	h_1	h_2	$\ldots$	h_α	$\ldots$	h_k
$\chi^{(1)}$	$f^{(1)}$	1	$\ldots$	1	$\ldots$	1
$\chi^{(2)}$	$f^{(2)}$	$\chi_2^{(2)}$	$\ldots$	$\chi_\alpha^{(2)}$	$\ldots$	$\chi_k^{(2)}$
$\vdots$	$\vdots$	$\vdots$		$\vdots$		$\vdots$
$\chi^{(i)}$	$f^{(i)}$	$\chi_2^{(i)}$	$\ldots$	$\chi_\alpha^{(i)}$	$\ldots$	$\chi_k^{(i)}$
$\vdots$	$\vdots$	$\vdots$		$\vdots$		$\vdots$
$\chi^{(k)}$	$f^{(k)}$	$\chi_2^{(k)}$	$\ldots$	$\chi_\alpha^{(k)}$	$\ldots$	$\chi_k^{(k)}$

Table 2.1.

Example. The group S_3 has three conjugacy classes, namely [13, Chapter VIII],

$$C_1 = \{1\}, C_2 = \{(12), (13), (23)\}, C_3 = \{(123), (132)\},$$

where $h_1 = 1$, $h_2 = 3$, $h_3 = 2$. Therefore S_3 has three simple characters: the trivial character $\chi^{(1)}(x)$, the alternating character $\chi^{(2)}(x)$ ($= \zeta(x)$, p. 4) and $\chi^{(3)}(x)$ ($= \phi(x)$, p. 42).

$$S_3:$$

	C_1	C_2	C_3
	1	3	2
$\chi^{(1)}$	1	1	1
$\chi^{(2)}$	1	-1	1
$\chi^{(3)}$	2	0	-1

Table 2.2.

We check that

$$(f^{(1)})^2 + (f^{(2)})^2 + (f^{(3)})^2 = 1^2 + 1^2 + 2^2 = 6,$$

which confirms that we have indeed found all the simple characters.

Let $A(x)$ be an arbitrary representation of degree m, not necessarily irreducible, and let $\phi(x)$ be the character of $A(x)$. Then

$$\phi(x) = \operatorname{tr} A(x) = sum\ of\ the\ latent\ roots\ (eigenvalues)\ of\ A(x).$$

If ε is a latent root of $A(x)$, then ε^n ($n = 0, \pm 1, \pm 2, \ldots$) is a latent root of $(A(x))^n = A(x^n)$. In a finite group of order g, each element satisfies the equation

$$x^g = 1.$$

Thus ε^g is a latent root of $A(x^g) = A(1) = I$, whence $\varepsilon^g = 1$, that is ε is a gth root of unity and therefore a complex number of modulus unity. Hence

$$\varepsilon^{-1} = \bar{\varepsilon},$$

where the bar denotes the conjugate complex.

If the latent roots of $A(x)$ are

$$\varepsilon_1(x), \varepsilon_2(x), \ldots, \varepsilon_m(x),$$

then those of $A(x^{-1})$ are

$$\overline{\varepsilon_1(x)}, \quad \overline{\varepsilon_2(x)}, \quad \ldots, \quad \overline{\varepsilon_m(x)}.$$

We conclude that, for any character,

$$\phi(x^{-1}) = \bar{\phi}(x), \qquad (2.37)$$

where the right-hand side is an abbreviation for $\overline{\phi(x)}$. This enables us to recast the expression for the inner product of two characters $\phi(x)$ and $\psi(x)$. When x lies in C_α, then $\phi(x) = \phi_\alpha$, $\psi(x) = \psi_\alpha$, $\psi(x^{-1}) = \bar{\psi}_\alpha$. Hence

$$\langle \phi, \psi \rangle = \frac{1}{g} \sum_{x \in G} \phi(x)\psi(x^{-1}) = \frac{1}{g} \sum_{\alpha=1}^{k} h_\alpha \phi_\alpha \bar{\psi}_\alpha. \qquad (2.38)$$

In particular, if χ and χ' are simple characters, the character relations (of the first kind, p. 39) state that

$$\frac{1}{g} \sum_{\alpha=1}^{k} h_\alpha \chi_\alpha \overline{\chi'_\alpha} = \begin{cases} 0 & \text{if } \chi \neq \chi' \\ 1 & \text{if } \chi = \chi' \end{cases}$$

Applying this result to all the rows of the character table we have that

$$\frac{1}{g} \sum_{\alpha=1}^{k} h_\alpha \chi_\alpha^{(i)} \overline{\chi_\alpha^{(j)}} = \delta_{ij} \qquad (2.39)$$

$(i, j = 1, 2, \ldots, k)$. We can express this information in a more compact form by introducing the $k \times k$ matrix

$$U = (u_{i\alpha}) = (\chi_\alpha^{(i)} \sqrt{(h_\alpha/g)}). \qquad (2.40)$$

Then (2.39) states that the rows of U form a system of unitary-orthogonal k-tuples. Thus U is a unitary matrix, that is

$$U\bar{U}' = I, \qquad (2.41)$$

where the dash denotes transposition. Now if U is a unitary matrix, so is U'; for, by (2.41), U and $\bar{U}'$ are inverse to each other and therefore commute. Hence $\bar{U}'U = I$ and, on taking the conjugate complex,

$$U'\bar{U} = I.$$

This means that the columns of U also form a unitary-orthogonal system, that is

$$\sum_{i=1}^{k} u_{i\alpha} \bar{u}_{i\beta} = \delta_{\alpha\beta},$$

or in terms of the characters

$$\sum_{i=1}^{k} \chi_\alpha^{(i)} \overline{\chi_\beta^{(i)}} = \frac{g}{h_\alpha} \delta_{\alpha\beta}. \qquad (2.42)$$

These equations are called the *character relations of the second kind*.

When x ranges over the conjugacy class C_α, then x^{-1} ranges over the conjugacy class $C_{\alpha'}$, whose elements are in one-to-one correspondence with those of C_α. We say that the classes C_α and $C_{\alpha'}$ are *inverse* to each other. It may happen that $C_\alpha = C_{\alpha'}$, in which case C_α is termed *self-inverse*. In any event,

$$h_{\alpha'} = h_\alpha,$$

and, for each character $\phi(x)$,

$$\phi_{\alpha'} = \overline{\phi_\alpha}.$$

Thus the character relations of the second kind can also be written

$$\sum_{i=1}^{k} \chi_\alpha^{(i)} \chi_{\beta'}^{(i)} = \frac{g}{h_\alpha} \delta_{\alpha\beta}. \tag{2.43}$$

Together with any representation $A(x)$ we have the contragredient representation (p. 16)

$$A^\dagger(x) = A'(x^{-1}).$$

If $\phi(x)$ and $\phi^\dagger(x)$ are the characters of $A(x)$ and $A^\dagger(x)$ respectively, we have by (2.37) that

$$\phi^\dagger(x) = \bar{\phi}(x).$$

Evidently, A and $A^\dagger$ are equivalent if and only if ϕ is real-valued.

In particular, if χ is a simple character, so is $\bar{\chi}$, because

$$\langle \bar{\chi}, \bar{\chi} \rangle = \langle \chi, \chi \rangle = 1.$$

2.4. Finite Abelian groups

The problem of the character table can be solved easily when G is a finite Abelian group. In this case each element forms a conjugacy class by itself so that $k = g$, and (2.26) becomes

$$\sum_{i=1}^{g} (f^{(i)})^2 = g.$$

This implies that each $f^{(i)}$ is equal to unity. Hence all absolutely irreducible representations of a finite Abelian group are linear. The one-rowed matrix $F^{(i)}(x)$ may then be identified with its sole coefficient and hence with its character, that is

$$F^{(i)}(x) = \chi^{(i)}(x),$$

and

$$\chi^{(i)}(xy) = \chi^{(i)}(x)\chi^{(i)}(y). \tag{2.44}$$

These *linear* characters are readily found, as we shall now demonstrate.

In the first place suppose that $G = Z_n$, a cyclic group of order n. For the sake of simplicity we shall temporarily use the notation

$$e(x) = \exp x = e^x.$$

Let

$$\varepsilon_r = e(2\pi i r/n)$$

be any nth root of unity, where, for the moment, r is a fixed integer satisfying $0 \leqslant r < n$. If

$$Z_n = \mathrm{gp}\{z\}, \qquad z^n = 1,$$

we define, for an arbitrary element $z^s \in Z_n$ $(s = 0, 1, \ldots, n-1)$,

$$\chi^{(r)}(z^s) = \varepsilon_r^s = e(2\pi i r s/n). \tag{2.45}$$

This is clearly a linear character of Z_n, because

$$\chi^{(r)}(z^s)\chi^{(r)}(z^t) = \varepsilon_r^{s+t} = \chi^{(r)}(z^{s+t}).$$

Distinct values of r yield distinct characters (2.45). As r ranges from 0 to $n-1$, we obtain n characters, as required. Contrary to our usual convention the trivial character will here be denoted by $\chi^{(0)}$. The complete character table of Z_n is then summarised by the equations

$$\chi^{(r)}(z^s) = e(2\pi i r s/n) \quad (r, s = 0, 1, \ldots, n-1). \tag{2.46}$$

The orthogonality relations

$$\langle \chi^{(i)}, \chi^{(j)} \rangle = \delta_{ij} \quad (i, j = 0, 1, \ldots, n-1)$$

could be checked; they follow from elementary identities for exponential functions.

Example. The group Z_3 consists of the elements

$$1, z, z^2 \quad (z^3 = 1).$$

Let $\omega = e(2\pi i/3)$. Then the character table is as in Table 2.3.

Z_3:

	1	z	z^2
$\chi^{(0)}$	1	1	1
$\chi^{(1)}$	1	ω	ω^2
$\chi^{(2)}$	1	ω^2	ω

Table 2.3.

For the general case of a finite Abelian group G we invoke the Basis Theorem, according to which G is the direct product of cyclic groups, say

$$G = \mathrm{gp}\{z_1\} \times \mathrm{gp}\{z_2\} \times \ldots \times \mathrm{gp}\{z_m\},$$

where z_μ is of order n_μ and

$$g = n_1 n_2 \ldots n_m. \tag{2.47}$$

An arbitrary element $x \in G$ is then uniquely expressed as

$$x = z_1^{a_1} z_2^{a_2} \ldots z_m^{a_m}, \tag{2.48}$$

where the exponents are subject to the conditions

$$0 \leqslant a_\mu < n_\mu \quad (\mu = 1, 2, \ldots, m).$$

In order to construct the simple characters of G we choose for each μ an n_μth root of unity

$$\varepsilon_\mu = \mathrm{e}(2\pi i r_\mu / n_\mu),$$

where r_μ is any integer satisfying

$$0 \leqslant r_\mu < n_\mu \quad (\mu = 1, 2, \ldots, m). \tag{2.49}$$

Corresponding to each m-tuple

$$r = [r_1, r_2, \ldots, r_m] \tag{2.50}$$

we define the function

$$\chi^{[r]}(x) = \mathrm{e}\left(2\pi i \sum_{\mu=1}^{m} a_\mu r_\mu / n_\mu\right), \tag{2.51}$$

x being given by (2.48). If

$$y = z_1^{b_1} z_2^{b_2} \ldots z_m^{b_m}$$

is another element of G, it is easy to verify that

$$\chi^{[r]}(x)\chi^{[r]}(y) = \chi^{[r]}(xy).$$

By (2.47) there are g m-tuples satisfying (2.49). Since distinct m-tuples correspond to distinct functions, the complete character table is presented in (2.51).

In every group the set of linear characters can be endowed with a group structure in which multiplication serves as composition. For if χ and χ' are functions with the property (2.44), then the function $\chi\chi'$ defined by

$$\chi\chi'(x) = \chi(x)\chi'(x)$$

54

also satisfies (2.44). In particular, the g simple characters of the Abelian group G form a multiplicative group in this sense, which is called the *dual group* of G and is often denoted by G^*.

In order to examine the relationship between G and G^* more fully, it is convenient to represent an element x of G by the m-tuple of exponents in (2.48), thus

$$x = (a_1, a_2, \ldots, a_m).$$

If

$$y = (b_1, b_2, \ldots, b_m)$$

is another element of G, then

$$xy = (a_1 + b_1, a_2 + b_2, \ldots, a_m + b_m),$$

provided that $a_\mu + b_\mu$ is replaced by its least non-negative remainder modulo n_μ.

Similarly, if $\chi^{[r]}$ and $\chi^{[s]}$ are two elements of G^*, then their product in G^* is given by

$$\chi^{[r]}\chi^{[s]} = \chi^{[r+s]},$$

where

$$r + s = [r_1 + s_1, r_2 + s_2, \ldots, r_m + s_m]$$

with the same rule regarding the reduction of $r_\mu + s_\mu$ modulo n_μ. Hence it is evident that the groups G and G^* are isomorphic, each being equivalent to the group of m-tuples with the composition described above.

The correspondence between G and G^* has been established with reference to a particular basis of G. We mention without proof that the intervention of a basis is unavoidable. Accordingly, the relationship

$$G \cong G^*$$

is termed a non-natural isomorphism.

The character theory for finite Abelian groups is summarised as follows:

Theorem 2.4. *Suppose that G is a direct product of m cyclic groups of orders $n_1, n_2, \ldots, n_m$ respectively, so that*

$$g = |G| = n_1 n_2 \ldots n_m.$$

Then a typical element of G can be represented by the m-tuple

$$x = (a_1, a_2, \ldots, a_m)$$

where $0 \leqslant a_\mu < n_\mu$. *Composition of elements is carried out by addition of components followed by reduction of the μth component to its least non-negative remainder modulo $n_\mu (\mu = 1, 2, \ldots, m)$.*

Corresponding to each m-tuple

$$r = [r_1, r_2, \ldots, r_m],$$

where $0 \leqslant r_\mu < n_\mu$, there exists a linear character

$$\chi^{[r]}(x) = e\left(2\pi i \sum_{\mu=1}^{m} a_\mu r_\mu / n_\mu\right)$$

of G, and all g characters are obtained in this way. The characters form a multiplicative group G^* which is isomorphic to G by virtue of the correspondence

$$(a_1, a_2, \ldots, a_m) \leftrightarrow [a_1, a_2, \ldots, a_m].$$

Example. The four-group is given by

$$V = \mathrm{gp}\{x\} \times \mathrm{gp}\{y\},$$

where $x^2 = y^2 = 1$, $xy = yx$. In order to obtain the characters of V we assign to x and y in turn the numbers 1 and -1, which then determines the character value of xy. Thus Table 2.4 is obtained.

V:

	1	x	y	xy
$\chi^{(1)}$	1	1	1	1
$\chi^{(2)}$	1	-1	1	-1
$\chi^{(3)}$	1	1	-1	-1
$\chi^{(4)}$	1	-1	-1	1

Table 2.4.

We conclude this section with an application to matrix groups.

Theorem 2.5. *Suppose that the $m \times m$ matrices over $\mathbb{C}$*

$$A_1(=I), A_2, \ldots, A_g$$

form an Abelian group under multiplication. Then these matrices can be simultaneously diagonalised, that is, there exists a non-singular matrix T such that

$$T^{-1} A_i T = \mathrm{diag}(\alpha_1^{(i)}, \alpha_2^{(i)}, \ldots, \alpha_m^{(i)}) \quad (i = 1, 2, \ldots, g).$$

56

Proof. A matrix group may be regarded as its own representation and is therefore equivalent to diagonally arranged blocks of irreducible representations. In the case of an Abelian group all irreducible representations are linear, which amounts to the possibility of simultaneous diagonalisation.

Corollary. *Every periodic matrix, that is a matrix satisfying*

$$A^p = I,$$

for some positive integer p, can be diagonalised.

Proof. We may assume that $p > 1$. The matrices

$$I, A, A^2, \ldots, A^{p-1}$$

form a cyclic group of order p. Hence there exists a non-singular matrix T such that $T^{-1}AT$ is a diagonal matrix.

2.5. The lifting process

The search for representations of a given group G is often facilitated by a knowledge about subgroups and quotient groups of G. The latter leads to a particularly simple procedure, which we shall develop in this section. Let $g = |G|$ and let

$$N : u_1(=1), u_2, \ldots, u_n$$

be a normal subgroup of G of order n. We recall that the elements of the quotient (factor) group G/N are the cosets Nx $(x \in G)$, of which exactly g/n are distinct. The law of composition in G/N is given by

$$NxNy = Nxy,$$

and the identity of G/N is N. Also $Nx = Ny$ if and only if $y = ux$, where $u \in N$. In particular, $Nu = N$.

Suppose that $A_0(Nx)$ is a representation (reducible or not) of G/N having degree m. Then

$$A_0(Nx)A_0(Ny) = A_0(Nxy),$$

$$A_0(N) = I_m.$$

Let

$$\phi_0(Nx) = \operatorname{tr} A_0(Nx)$$

be the character of $A_0(Nx)$.

57

We now use this representation to construct a representation A of G by defining

$$A(x) = A_0(Nx) \quad (x \in G).$$

This is indeed a representation of G, because

$$A(x)A(y) = A_0(Nx)A_0(Ny) = A_0(Nxy) = A(xy).$$

The character of $A(x)$ is given by

$$\phi(x) = \phi_0(Nx).$$

We note that if $u \in N$, then

$$A(u) = A_0(N) = I_m, \quad \phi(u) = m.$$

Since $A(x)$ consists of the same matrices as $A_0(Nx)$, it is clear that $A(x)$ is irreducible if and only if $A_0(Nx)$ is irreducible. Thus we have the following result:

Theorem 2.6. (*Lifting process*). *Let N be a normal subgroup of G and suppose that $A_0(Nx)$ is a representation of degree m of the group G/N. Then*

$$A(x) = A_0(Nx)$$

defines a representation of G, 'lifted' from G/N. If $\phi_0(Nx)$ is the character of $A_0(Nx)$, then

$$\phi(x) = \phi_0(Nx)$$

is the 'lifted' character of $A(x)$. Also, if $u \in N$,

$$A(u) = I_m, \quad \phi(u) = m = \phi(1).$$

The lifting operation preserves irreducibility.

We now ask the opposite question: how can we tell whether a given representation of degree m has been lifted from the representation $A_0(Nx)$ of a suitable factor group G/N? This is related to the problem of finding the kernel of the map

$$x \to A(x).$$

Let this kernel be denoted by M. Then if $u \in M$, $A(u) = I_m$ and hence

$$\phi(u) = m. \tag{2.52}$$

Conversely, suppose that an element u of G satisfies (2.52). Since $A(u)$ is a periodic matrix, the corollary to Theorem 2.5 implies that

$$A(u) \sim \mathrm{diag}(\varepsilon_1, \varepsilon_2, \ldots, \varepsilon_m), \tag{2.53}$$

where each ε_i is a root of unity. Taking the trace we find that

$$m = \varepsilon_1 + \varepsilon_2 + \ldots + \varepsilon_m.$$

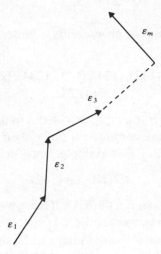

Figure 1.

We have that (see Figure 1)

$$m = |\varepsilon_1 + \varepsilon_2 + \ldots + \varepsilon_m| \leqslant |\varepsilon_1| + |\varepsilon_2| + \ldots + |\varepsilon_m| = m,$$

and equality can hold only if $\varepsilon_1 = \varepsilon_2 = \ldots = \varepsilon_m = \varepsilon$, say. Hence $m = m\varepsilon$, $\varepsilon = 1$, which means that $A(u) = I$. In other words, the condition (2.52) completely characterises the kernel.

We shall now construct a representation of G/M by putting

$$A_0(Mx) = A(x). \tag{2.54}$$

We must, however, verify that this representation is well-defined; indeed if $Mx = My$, then $y = ux$, where $u \in M$, and by (2.54),

$$A_0(My) = A(y) = A(ux) = A(u)A(x) = IA(x) = A_0(Mx),$$

as required.

The representation $A(x)$ is faithful if and only if $\phi(x) = \phi(1)$ implies that $x = 1$.

Summarising this discussion we have the following result.

Theorem 2.7. *If A is a representation of G with character ϕ, then those elements u which satisfy $\phi(u) = \phi(1)$ form a normal subgroup M, which is the kernel of A. Then $A_0(Mx) = A(x)$ is a well-defined representation of G/M, and we may regard A as having been lifted from A_0.*

In particular, we can discover normal subgroups of G by running through each row of the character table and gathering together all those elements u for which $\chi(u) = \chi(1)$.

Example 1. The alternating group A_4 is of order 12 and consists of the following elements:

$$1, (123), (132), (124), (142), (134), (143), (234), (243), (12)(34), (13)(24),$$
$$(14)(23).$$

First, we must split A_4 into conjugacy classes. In the full symmetric group S_4 two permutations are conjugate if and only if they possess the same cycle pattern [13, p. 131]. For example, consider

$$t^{-1}(123)t = (132), \tag{2.55}$$

where $t = (23)$, which does not lie in A_4. If there were such a permutation t in A_4, it would have to leave the object 4 fixed. Hence t would have to be 1 or (123) or (132), none of which satisfy (2.55), since they commute with (123). We must therefore conclude that (123) and (132) belong to distinct conjugacy classes of A_4. The same observation applies to other pairs of 3-cycles that involve the same set of objects. On the other hand (12)(34), (13)(24) and (14)(23) remain conjugate in A_4, for example

$$(234)^{-1}(12)(34)(234) = (13)(24).$$

Thus it is found that the conjugacy classes of A_4 are as follows:

$$C_1 = \{1\}$$
$$C_2 = \{(123), (142), (134), (243)\}$$
$$C_3 = \{(132), (124), (143), (234)\}$$
$$C_4 = \{(12)(34), (13)(24), (14)(23)\}.$$

We infer that A_4 possesses four irreducible representations. It so happens that $V = C_1 \cup C_4$ is a normal subgroup of A_4 [13, p. 138]. The quotient group A_4/V is of order 3 and is therefore isomorphic to Z_3, whose character table was given on p. 53. According to Theorem 2.6 we can lift the three characters of Z_3, all of which are linear, to obtain three characters $\chi^{(1)}, \chi^{(2)}, \chi^{(3)}$ of A_4, the kernel being V. It remains to find $\chi^{(4)}$. Since

$$(f^{(1)})^2 + (f^{(2)})^2 + (f^{(3)})^2 + (f^{(4)})^2 = 12,$$

and $f^{(1)} = f^{(2)} = f^{(3)} = 1$, we deduce that $f^{(4)} = 3$. For the moment denote the values of $\chi^{(4)}$ for the classes C_2, C_3, C_4 by α, β, γ respectively.

Thus the character table, so far, is of the form of Table 2.5(a).

A_4:

	C_1	C_2	C_3	C_4
h	1	4	4	3
$\chi^{(1)}$	1	1	1	1
$\chi^{(2)}$	1	ω	ω^2	1
$\chi^{(3)}$	1	ω^2	ω	1
$\chi^{(4)}$	3	α	β	γ

Table 2.5(a).

The values of α, β and γ are readily derived from the orthogonality of the columns (character relations of the second kind). Thus it is found that $\alpha = \beta = 0$, because $1 + \omega + \omega^2 = 0$ and $3\gamma + 3 = 0$, whence $\gamma = -1$. Hence the complete character table is as given in Table 2.5(b).

A_4:

	C_1	C_2	C_3	C_4
h	1	4	4	3
$\chi^{(1)}$	1	1	1	1
$\chi^{(2)}$	1	ω	ω^2	1
$\chi^{(3)}$	1	ω^2	ω	1
$\chi^{(4)}$	3	0	0	-1

Table 2.5(b).

Example 2. The dihedral group D_4 of order 8 is defined by

$$a^4 = b^2 = 1, \qquad b^{-1}ab = a^3.$$

We note that $b^{-1}a^2b = a^6 = a^2$. Hence a^2 lies in the centre, as it commutes with both a and b. In fact, the centre is $\{1, a^2\}$. Thus 1 and a^2 form conjugacy classes by themselves, the other conjugacy classes being

$$\{a, a^3\}, \{b, a^2b\}, \{ab, a^3b\}.$$

Since there are five conjugacy classes in all, the group D_4 possesses five irreducible representations.

Evidently, the subgroup $N = 1 \cup a^2$ is normal in D_4. The quotient group D_4/N consists of the elements

$$N, \qquad Na, \qquad Nb, \qquad Nab$$

and is isomorphic to the four-group, whose character table is displayed on p. 56. The four characters of D_4/N are lifted to yield linear characters $\chi^{(1)}, \chi^{(2)}, \chi^{(3)}, \chi^{(4)}$ of D_4. It remains to determine $\chi^{(5)}$. Since $f^{(1)} = f^{(2)} = f^{(3)} = f^{(4)} = 1$ and

$$\sum_{i=1}^{5} (f^{(i)})^2 = 8,$$

it follows that $f^{(5)} = 2$. The other values of $\chi^{(5)}$ are deduced from the fact that the columns of the character table are mutually orthogonal. The result is in Table 2.6.

D_4 or Q: $\qquad a^3b$

C	1	a^2	$\{a, a^3\}$	$\{b, a^2b\}$	$\{ab, a^2b\}$
h	1	1	2	2	2
$\chi^{(1)}$	1	1	1	1	1
$\chi^{(2)}$	1	1	1	-1	-1
$\chi^{(3)}$	1	1	-1	1	-1
$\chi^{(4)}$	1	1	-1	-1	1
$\chi^{(5)}$	2	-2	0	0	0

Table 2.6.

Example 3. The quaternion group Q of order 8 is defined by

$$a^4 = 1, \qquad a^2 = b^2, \qquad b^{-1}ab = a^3.$$

(It is not isomorphic to D_4.) The group $N = \{1, a^2\}$ is a normal subgroup of Q, and in fact is its centre. Also the group Q/N consists of the cosets

$$N, \qquad Na, \qquad Nb, \qquad Nab$$

and is isomorphic to the four-group. Hence the same arguments apply for the construction of the character table of Q as in the case of D_4. Thus these two groups have the same character tables. This is a somewhat disappointing discovery, as it dashes all hope of using the character table to discriminate between non-isomorphic groups.

2.6. Linear characters

The problem of finding all the linear characters of G is related to the commutator group (or derived group) G'. We recall that

$$G' = \mathrm{gp}\{x^{-1}y^{-1}xy \,|\, x, y \in G\}.$$

It is known that G' is normal in G and that G/G' is Abelian. Also, G' is the least subgroup with these properties; that is, if N is a normal subgroup such that G/N is Abelian, then $G' \leqslant N$ [**13**, p. 74].

Since G/G' is Abelian, all its irreducible characters are linear. Suppose that $|G/G'| = l$, and let

$$\lambda_0^{(1)}, \lambda_0^{(2)}, \ldots, \lambda_0^{(l)} \tag{2.56}$$

be a complete set of irreducible characters of G/G'. Then we obtain l characters of G by lifting from G/G', namely

$$\lambda^{(i)}(x) = \lambda_0^{(i)}(G'x) \quad (i = 1, 2, \ldots, l). \tag{2.57}$$

Conversely, we shall show that all linear characters of G are obtained in this way. Thus, let $\mu(x)$ be a linear character of G. For any commutator $a^{-1}b^{-1}ab$ we find that

$$\mu(a^{-1}b^{-1}ab) = \mu(a^{-1})\mu(b^{-1})\mu(a)\mu(b) = 1.$$

Since the commutators generate G', it follows that

$$\mu(u) = 1, \quad \text{if } u \in G',$$

that is G' lies in the kernel of every linear representation of G. We can now define a representation μ_0 of G/G' by

$$\mu_0(G'x) = \mu(x). \tag{2.58}$$

This is legitimate, because if $G'x = G'y$ so that $y = ux$, $u \in G'$, we have that

$$\mu(y) = \mu(ux) = \mu(u)\mu(x) = \mu(x),$$

as required. We deduce that μ_0 must be one of the representations of G/G' listed in (2.56), and it follows from (2.58) that μ is one of the representations of G defined in (2.57). Summarising we have

Theorem 2.8. *The number of linear characters of G is equal to $|G/G'|$. There is a one-to-one correspondence between the simple characters λ_0 of G/G', which are necessarily linear, and the linear characters of G, given by*

$$\lambda(x) = \lambda_0(G'x) \quad (x \in G).$$

All linear characters of G take the value unity on G'.

Example 4. According to Table 2.5(b), the group A_4 has three linear characters. Thus $|A_4/A_4'| = 3$, whence $|A_4'| = 4$. Now V is a normal subgroup of A_4 such that A_4/V is Abelian. By the minimal property of A_4', we have that $V \geqslant A_4'$. Since $|V| = |A_4'|$, it follows that $V = A_4'$.

We conclude this section with an observation that is sometimes useful in our search for representations. Let $A(x)$ be any representation of G, irreducible or not, and suppose that $\lambda(x)$ is a linear character. Then

$$B(x) = \lambda(x)A(x)$$

is also a representation, since it is easily verified that

$$B(xy) = B(x)B(y).$$

If $A(x)$ has character $\phi(x)$, then $B(x)$ has character $\lambda(x)\phi(x)$. This applies, in particular, when $A(x)$ is irreducible. Thus we have

Proposition 2.2. *If $\lambda(x)$ is a linear character, then together with every irreducible character, the character table also contains $\lambda(x)\chi(x)$. For, since $\lambda(x)\lambda(x^{-1}) = 1$, it is obvious from (2.5) that*

$$\langle \lambda\chi, \lambda\chi \rangle = \langle \chi, \chi \rangle = 1.$$

However, it may happen that $\chi(x) = \lambda(x)\chi(x)$ for all $x \in G$, in which case the method produces no new character.

Exercises

1. Let $A = (a_{ij}(x))$ be an absolutely irreducible representation of a finite group G. Prove that

$$\sum_{y \in G} a_{ip}(y^{-1})a_{ps}(yx) = (g/f)a_{is}(x) \quad (x \in G),$$

where $g = |G|$ and $f = \deg A$.

2. Let A be an absolutely irreducible representation of degree f of the finite group G. Prove that if the (constant) $f \times f$ matrix K has the property that

$$\mathrm{tr}\{KA(x)\} = 0, \quad \text{for all } x \text{ in } G,$$

then $K = 0$.

3. (i) Show that the permutation representation of S_3 induced by the subgroup H: 1, (12) is isomorphic to S_3 (see p. 2).

(ii) Show that the permutation representation of S_3 induced by the subgroup K: 1, (123), (132) is equivalent to

$$\begin{pmatrix} 1 & 0 \\ 0 & \zeta(x) \end{pmatrix} \quad (x \in S_3),$$

where ζ is the alternating character of S_3 (p. 50).

4. Let $X = (\chi_\alpha^{(i)})$ $(i, \alpha = 1, 2, \ldots, k)$ be the matrix comprising the character table of a group. Show that X is non-singular.

Prove that the number of real-valued characters is equal to the number of self-inverse conjugacy classes.

64

5. Show that in a group G of odd order no element other than 1 is conjugate to its inverse.

Prove that if $u \in G$ and $u \neq 1$, there exists at least one simple character χ such that $\chi(u)$ is not real.

6. Prove that, in a group of odd order, the trivial character is the only character that is real-valued.

7. Let $A(x)$ and $B(x)$ be two matrix representations over $\mathbb{C}$ of a finite group G. Suppose that for each $x \in G$ there exists a non-singular matrix $P(x)$ such that

$$\{P(x)\}^{-1}A(x)P(x) = B(x).$$

Prove that there exists a non-singular matrix T, independent of x, such that

$$T^{-1}A(x)T = B(x).$$

8. Prove that, if a finite group has a faithful absolutely irreducible representation, then its centre is either trivial or else cyclic.

9. Let $G = \mathrm{gp}\{a, b\}$ be the *dicyclic* group of order 12 defined by the relations

$$a^6 = 1, \qquad a^3 = (ab)^2 = b^2.$$

Show that the derived group and the centre are $G' = \{1, a^2, a^4\}$ and $Z = \{1, a^3\}$ respectively. By lifting characters from G/G' and G/Z, or otherwise, establish the character table given in Table 2.7.

	1	a^3	(a, a^5)	(a^2, a^4)	(b, a^2b, a^4b)	(ab, a^3b, a^5b)
$\chi^{(1)}$	1	1	1	1	1	1
$\chi^{(2)}$	1	-1	-1	1	i	$-i$
$\chi^{(3)}$	1	1	1	1	-1	-1
$\chi^{(4)}$	1	-1	-1	1	$-i$	i
$\chi^{(5)}$	2	2	-1	-1	0	0
$\chi^{(6)}$	2	-2	1	-1	0	0

Table 2.7.

10. The *dihedral group* D_{2m} of order $4m$ is defined by the relations

$$a^{2m} = b^2 = 1, \qquad b^{-1}ab = a^{-1}.$$

Show that there are $m+3$ conjugacy classes, namely $C_0 = (1)$, $C_j = (a^j, a^{2m-j})$ ($j = 1, 2, \ldots, m-1$), $C_m = (a^m)$, $C_{m+1} = (b, a^2b, \ldots, a^{2m-2}b)$, $C_{m+2} = (ab, a^3b, \ldots, a^{2m-1}b)$. Prove that $D'_{2m} = \mathrm{gp}\{a^2\}$. Put $\varepsilon = \exp(\pi i/m)$ and let t be an integer such that $1 \leq t \leq m-1$. Verify that the matrices

$$A_t = \begin{pmatrix} \varepsilon^t & 0 \\ 0 & \varepsilon^{-t} \end{pmatrix}, \qquad B_t = \begin{pmatrix} 0 & 1 \\ 1 & 0 \end{pmatrix}$$

furnish an irreducible representation over $\mathbb{C}$ by virtue of the map $a \to A_t$, $b \to B_t$. Deduce that the character table of D_{2m} consists of four linear characters

$$\lambda^{(r,s)}(a^\alpha b^\beta) = (-1)^{\alpha r + \beta s} \quad (r, s = 0, 1)$$

65

and $m - 1$ two-dimensional characters

$$\chi^{(t)}(a^\alpha) = 2 \cos \alpha t\phi \quad (\alpha = 0, 1, \ldots, 2m - 1)$$
$$\chi^{(t)}(a^\alpha b) = 0,$$

where $\phi = \pi/m$.

11. The *dihedral group* D_{2m+1} of order $4m + 2$ is defined by the relations

$$a^{2m+1} = b^2 = 1, \, b^{-1}ab = a^{-1}.$$

Show that there are $m + 2$ conjugacy classes, namely $C_0 = (1)$, $C_j = (a^j, a^{2m+1-j})$ $(j = 1, 2, \ldots, m)$, $C_{m+1} = (b, ab, \ldots, a^{2m}b)$. Prove that $D'_{2m+1} = \mathrm{gp}\{a\}$. Put $\varepsilon = \exp(2\pi i/2m + 1)$ and let t be an integer satisfying $1 \leqslant t \leqslant m$. Verify that the matrices

$$A_t = \begin{pmatrix} \varepsilon^t & 0 \\ 0 & \varepsilon^{-t} \end{pmatrix}, \qquad B_t = \begin{pmatrix} 0 & 1 \\ 1 & 0 \end{pmatrix}$$

furnish an irreducible representation over $\mathbb{C}$ by virtue of the map $a \to A_t, b \to B_t$. Deduce that the character table of D_{2m+1} consists of two linear characters

$$\lambda^{(r)}(a^\alpha b^\beta) = (-1)^{\beta r} \quad (r = 0, 1)$$

and m two-dimensional characters

$$\chi^{(t)}(a^\alpha) = 2 \cos \alpha t\phi \quad (\alpha = 0, 1, \ldots, 2m)$$
$$\chi^{(t)}(a^\alpha b) = 0,$$

where $\phi = 2\pi/2m + 1$.

12. Show that the matrix representations used in the two preceding examples can be brought into real form; that is, there exists a matrix P such that $P^{-1}A_tP$ and $P^{-1}B_tP$ are real.

13. Let H be a normal subgroup of G such that $G/H = \mathrm{gp}\{Ht\}$ is cyclic of order m, where $t^m = v \in H$. Suppose that θ is a linear character of H with the properties that

(i) $\theta(t^r u t^{-r}) = \theta(u) \quad (u \in H; r = 0, 1, \ldots, m - 1),$

(ii) $\theta(v) = 1.$

Prove that if ε is an mth root of unity, then

$$\Theta(ut^r) = \theta(u)\varepsilon^r$$

is a linear character of G.

14. Prove that if $A(x)$ is a representation of a finite group G such that $\det A(u) \neq 1$ for at least one element u of G, then $[G:G'] > 1$.

15. Let $G: x_1, x_2, \ldots, x_g$ be a group of order g, and let p and q run through the elements of G in this order. With each x of G associate an indeterminate ξ_x. The $g \times g$ matrix

$$\Gamma = (\xi_{p^{-1}q}),$$

in which p and q specify the rows and columns respectively, is called the *group matrix* of G. Write

$$\Gamma = \sum_x R(x)\xi_x,$$

where $R(x)$ is the matrix which embodies the coefficients of ξ_x in Γ. Prove that the matrices $R(x)$ are identical with those of the regular representation, referred to the natural basis of G_C (p. 44).

16. Using the notation of the preceding exercise we call $\det(\xi_{p^{-1}q})$ the *group determinant* of G. Prove that

$$\det(\xi_{p^{-1}q}) = \prod_{i=1}^{k} \Phi_i^{f^{(i)}},$$

where

$$\Phi_i = \det\left(\sum_x F^{(i)}(x)\xi_x\right).$$

In particular, when G is a finite Abelian group with linear characters $\lambda^{(i)}(x)$ $(i = 1, 2, \ldots, g)$, then

$$\det(\xi_{p^{-1}q}) = \prod_{i=1}^{g}\left(\sum_x \lambda^{(i)}(x)\xi_x\right).$$

Historical remark. The case of an Abelian group having been known for some time, Dedekind wrote to Frobenius in April 1896 posing the problem of factorising the group determinant of a non-Abelian group. In answer to this question Frobenius created the theory of group characters, which he published in the same year.

17. Establish the formula for the circulant determinant:

$$\begin{vmatrix} \xi_0 & \xi_1 & \xi_2 & \cdots & \xi_{n-1} \\ \xi_{n-1} & \xi_0 & \xi_1 & \cdots & \xi_{n-2} \\ \cdots & \cdots & & \cdots & \\ \xi_1 & \xi_2 & \xi_3 & \cdots & \xi_0 \end{vmatrix} = \prod_{r=0}^{n-1} (\xi_0 + \varepsilon^r \xi_1 + \varepsilon^{2r}\xi_2 + \ldots + \varepsilon^{(n-1)r}\xi_{n-1}),$$

where $\varepsilon = \exp(2\pi i/n)$.

18. Let V: 1, a, b, ab be the four-group defined by the relations $a^2 = b^2 = 1$, $ab = ba$. Show that each of the elements

$$u_1 = 1 + a + b + ab, \qquad u_2 = 1 - a + b - ab,$$
$$u_3 = 1 + a - b - ab, \qquad u_4 = 1 - a - b + ab$$

of the group algebra V_C (p. 44) generates a one-dimensional V-module.

67

19. Let $A_1, A_2, \ldots, A_n$ be a set of matrices over $\mathbb{C}$, at least one of which is non-singular. Prove that there exists a non-singular matrix P such that

$$\sum_{i=1}^{n} A_i \bar{A}_i' = P\bar{P}'.$$

Also show that, if each A_i is real, then P may be taken to be real.

20. Let $A(x)$ be a representation of a finite group G and put

$$H = \sum_{y \in G} A(y)\bar{A}'(y).$$

Prove that

$$A(x)H\bar{A}'(x) = H \quad (x \in G).$$

Deduce that $A(x)$ is equivalent to a unitary representation. Show also that, if each $A(x)$ is real, then $B(x)$ may be taken to be real orthogonal.

21. Use Exercises 7 of Chapter 1 and 20 of Chapter 2 to give an alternative proof of Maschke's Theorem, valid for the field $\mathbb{C}$.

3

INDUCED CHARACTERS

3.1. Induced representations

With a brilliant idea, Frobenius succeeded in constructing representations of a group from those of an arbitrary subgroup.

Let G be a group of order g, and let H be a subgroup of order h and index n $(= g/h)$. Suppose that $B(u)(u \in H)$ is a representation of H of degree q, thus

$$B(u)B(v) = B(uv) \quad (u, v \in H). \tag{3.1}$$

We formally extend the matrix function B to the whole of G by putting

$$B(x) = 0, \text{ if } x \notin H. \tag{3.2}$$

Of course, this convention does not turn $B(x)$ into a representation of G, if only for the obvious reason that, by their very definition, all representations consist of non-singular matrices. A more subtle procedure is required. Let

$$G = Ht_1 \cup Ht_2 \cup \ldots \cup Ht_n$$

be a decomposition of G into right cosets of H, that is, let $t_1, t_2, \ldots, t_n$ be a right *transversal* of H in G. For every element x of G we define a matrix $A(x)$ of degree qn as an $n \times n$ array of blocks, each of degree q, as follows:

$$A(x) = \begin{pmatrix} B(t_1 x t_1^{-1}) & B(t_1 x t_2^{-1}) & \ldots & B(t_1 x t_n^{-1}) \\ B(t_2 x t_1^{-1}) & B(t_2 x t_2^{-1}) & \ldots & B(t_2 x t_n^{-1}) \\ \ldots & \ldots & & \ldots \\ B(t_n x t_1^{-1}) & B(t_n x t_2^{-1}) & \ldots & B(t_n x t_n^{-1}) \end{pmatrix},$$

or, more briefly,

$$A(x) = \left(B(t_i x t_j^{-1}) \right). \tag{3.3}$$

The fundamental discovery of Frobenius was the fact that $A(x)$ is indeed a representation of G, that is

$$A(x)A(y) = A(xy) \quad (x, y \in G). \tag{3.4}$$

69

Comparing the (i, j)th block on both sides we see that (3.4) is equivalent to the assertion that, for all fixed i, j, x and y,

$$\sum_{r=1}^{n} B(t_i x t_r^{-1}) B(t_r y t_j^{-1}) = B(t_i x y t_j^{-1}). \tag{3.5}$$

Two cases arise according to the nature of the right-hand side of (3.5).

(i) Suppose that $t_i x y t_j^{-1} \notin H$. Hence by (3.2) the right-hand side of (3.5) is the zero matrix. We claim that, for each r ($1 \le r \le n$), either $t_i x t_r^{-1} \notin H$ or $t_r y t_j^{-1} \notin H$. For otherwise we should have that

$$(t_i x t_r^{-1})(t_r y t_j^{-1}) = t_i x y t_j^{-1} \in H,$$

which contradicts our present hypothesis. Hence each term on the left-hand side of (3.5) is zero, and the two sides agree.

(ii) Suppose that $v = t_i x y t_j^{-1} \in H$. The element $t_i x$ belongs to exactly one right coset, say

$$t_i x \in H t_s \text{ so that } u = t_i x t_s^{-1} \in H.$$

Hence, if $r \ne s$, we have that $t_i x t_r^{-1} \notin H$. It follows that the sum on the left of (3.5) reduces to the single term in which $r = s$. This term is surely non-zero because

$$t_s y t_j^{-1} = u^{-1} v \in H,$$

and (3.5) is equivalent to

$$B(u) B(u^{-1} v) = B(v),$$

which is true by virtue of (3.1). This completes the proof of (3.4).

Our argument has shown that, for a fixed value of i, exactly one block in (3.3) is non-zero. Similarly, for a fixed value of j, precisely one block is non-zero. This happens when i is such that $t_i x t_j^{-1} \in H$, that is when $t_i \in H t_j x^{-1}$. Thus $A(x)$ can be described as a generalisation of a permutation matrix, the units in such a matrix having been replaced by blocks B. The representation $A(x)$ of G is said to have been *induced* by the representation $B(u)$ of H.

Let $\phi(u)$ be the character of $B(u)$. In conformity with (3.2) we define

$$\phi(x) = 0, \quad \text{if } x \notin H. \tag{3.6}$$

The character of $A(x)$, which is called the *induced character* of ϕ, will be denoted by ϕ^G. Thus

$$\phi^G(x) = \operatorname{tr} A(x) = \sum_{i=1}^{n} \phi(t_i x t_i^{-1}). \tag{3.7}$$

In the construction of $A(x)$ we employed a particular transversal, but we shall now demonstrate that this choice does not materially affect the result. For $s_1, s_2, \ldots, s_n$ is another transversal if and only if

$$s_i = u_i t_i \quad (i = 1, 2, \ldots, n), \tag{3.8}$$

where $u_1, u_2, \ldots, u_n$ are suitable elements of H. Now

$$s_i x s_i^{-1} = u_i t_i x t_i^{-1} u_i^{-1}.$$

If $s_i x s_i^{-1}$ lies in H, so does $t_i x t_i^{-1}$ and, since these elements are conjugate in H,

$$\phi(s_i x s_i^{-1}) = \phi(t_i x t_i^{-1}). \tag{3.9}$$

But if $s_i x s_i^{-1}$ does not belong to H, then neither does $t_i x t_i^{-1}$, and (3.9) is trivially satisfied because both sides are zero. On summing (3.9) over i we find that

$$\phi^G(x) = \sum_{i=1}^{n} \phi(s_i x s_i^{-1}),$$

as required.

There are alternative ways of writing the formula for the induced character. Let u be an arbitrary element of H and choose the transversal $s_i = u t_i$ ($i = 1, 2, \ldots, n$), that is put $u_1 = u_2 = \ldots = u_n = u$ in (3.8). Then

$$\phi^G(x) = \sum_{i=1}^{n} \phi(u t_i x t_i^{-1} u^{-1}).$$

Summing over u in H we note that the left-hand side is multiplied by h, while the products $u t_i$ run over G exactly once. Hence

$$\phi^G(x) = \frac{1}{h} \sum_{y \in G} \phi(y x y^{-1}). \tag{3.10}$$

This result can be cast into yet another form if we consider elements of a particular conjugacy class C_α. Thus if $x \in C_\alpha$ we put

$$\phi_\alpha^G = \phi^G(x).$$

There are h_α distinct conjugates of x, where $h_\alpha = |C_\alpha|$. As y runs through G, the element $y x y^{-1}$ is always conjugate with x in G, but these elements are not all distinct. In fact,

$$y_1 x y_1^{-1} = y_2 x y_2^{-1} \tag{3.11}$$

if and only if $y_1^{-1} y_2$ commutes with x. We recall that the set of elements of G which commute with x, is called the *centraliser* of x in G. This is a

71

subgroup of G which we shall here denote by Z_x. Hence (3.11) is equivalent to $y_1^{-1}y_2 \in Z_x$, that is, to $y_1 Z_x = y_2 Z_x$. Thus there is a one-to-one correspondence between the distinct conjugates of x and the (left) cosets of its centraliser in G. In particular, we have that

$$h_\alpha = [G : Z_x],$$

the index of Z_x in G. When y runs through G, the elements yxy^{-1} cover the conjugacy class C_α exactly g/h_α ($= |Z_x|$) times. Hence (3.10) becomes

$$\phi_\alpha^G = \frac{g}{hh_\alpha} \sum_{z \in C_\alpha} \phi(z) = \frac{n}{h_\alpha} \sum_{z \in C_\alpha} \phi(z), \tag{3.12}$$

where $n = g/h$. A further simplification is brought about by the fact that $\phi(z) = 0$ unless $z \in H$. Therefore, the summation in (3.12) may be restricted to $C_\alpha \cap H$, that is

$$\phi_\alpha^G = \frac{n}{h_\alpha} \sum_w \phi(w) \quad (w \in C_\alpha \cap H). \tag{3.13}$$

This formula is especially useful for practical computations.

It must be stressed that even when ϕ is a simple character of B, the induced character ϕ^G is, in general, a compound character of G. This is hardly surprising since the degree of ϕ^G is n times as large as the degree of ϕ.

By way of illustration, it is interesting to observe that the permutation representation

$$\sigma(x) = \begin{pmatrix} Ht_1 & Ht_2 & \dots & Ht_n \\ Ht_1 x & Ht_2 x & \dots & Ht_n x \end{pmatrix}$$

of G, which we mentioned in Chapter 1 (p. 2), is a special case of an induced representation, namely that which corresponds to the trivial representation of H. Let $\nu(x)$ be the character of $\sigma(x)$, viewed as a representation by permutation matrices. Then by (1.48), $\nu(x)$ is equal to the number of objects fixed under the action of x. Now $Ht_i x = Ht_i$ if and only if $t_i x t_i^{-1} \in H$. Denoting the trivial character of H, or its extension to G, by ξ, we have that $\xi(u) = 1$, if $u \in H$, and $\xi(x) = 0$ if $x \notin H$. Hence we can express $\nu(x)$ by the formula

$$\nu(x) = \sum_{i=1}^{n} \xi(t_i x t_i^{-1}).$$

Comparing this with (3.7) we find that

$$\nu = \xi^G.$$

As we have already remarked (p. 19), the permutation character is compound when $n > 1$. More generally, any character ψ, other than $\chi^{(1)}$, is compound if its values are non-negative real numbers. For then

$$c_1 = \langle \psi, \chi^{(1)} \rangle = \frac{1}{g} \sum_{\alpha=1}^{k} h_\alpha \psi_\alpha > 0,$$

and it follows that the Fourier expansion of ψ is of the form

$$\psi = c_1 \chi^{(1)} + c_2 \chi^{(2)} + \dots,$$

where c_1 is a positive integer. Hence

$$\psi - \chi^{(1)} = (c_1 - 1)\chi^{(1)} + c_2 \chi^{(2)} + \dots$$

is also a character, which may still be compound.

We record the result for future reference.

Proposition 3.1. *If G has a permutation representation $\pi(x)$ of degree greater than unity, then the function*

$$\phi(x) = (number\ of\ fixed\ points\ of\ \pi(x)) - 1 \qquad (3.14)$$

is a character of G, possibly compound.

3.2. The reciprocity law

The process of induction raises a character of a subgroup H to a character of the full group G. There is a trivial operation that works in the opposite direction: if $A(x)$ is a representation of G, then by restricting x to elements of H we obtain a representation $A(u)$ of H, because the equation

$$A(u)A(v) = A(uv)$$

clearly holds for all u and v in H. If ϕ is the character of A, we denote the *restricted* (or *deduced*) *character* by ϕ_H, that is

$$\phi_H(u) = \phi(u) \quad (u \in H).$$

Again, it must be pointed out that the process of restriction does not, in general, preserve irreducibility. Thus it may happen that, while $A(x)$ is an irreducible representation of G, the set of matrices $A(u)$ $(u \in H)$ is reducible over the ground field. When dealing with inner products we must now indicate which group is involved. Thus we shall use the notation

$$\langle \sigma, \rho \rangle_G = \frac{1}{g} \sum_{x \in G} \sigma(x) \bar{\rho}(x)$$

and analogously

$$\langle \sigma, \rho \rangle_H = \frac{1}{h} \sum_{u \in H} \sigma(u)\bar{\rho}(u).$$

The processes of induction and restriction are linked by a result which, though not particularly deep, furnishes us with a useful tool for handling group characters.

Theorem 3.1 (*Reciprocity Theorem of Frobenius*). *Let H be a subgroup of G. If ψ and ϕ are characters of H and G respectively, then*

$$\langle \psi^G, \phi \rangle_G = \langle \psi, \phi_H \rangle_H. \qquad (3.15)$$

Proof. Applying (3.10) to ψ we have that

$$\psi^G(x) = \frac{1}{h} \sum_{y \in G} \psi(yxy^{-1}),$$

where $\psi(z) = 0$ if $z \notin H$. Then

$$\langle \psi^G, \phi \rangle_G = (gh)^{-1} \sum_{x \in G} \sum_{y \in G} \psi(yxy^{-1})\bar{\phi}(x)$$

$$= (gh)^{-1} \sum_{y \in G} \sum_{x \in G} \psi(yxy^{-1})\bar{\phi}(x).$$

For fixed y let $z = yxy^{-1}$, that is $x = y^{-1}zy$. As x runs through G so does z. Hence we may use z as the summation variable in the inner sum and obtain that

$$\langle \psi^G, \phi \rangle_G = (gh)^{-1} \sum_{y \in G} \sum_{z \in G} \psi(z)\bar{\phi}(y^{-1}zy).$$

Since ϕ is a class function on G, we have $\phi(y^{-1}zy) = \phi(z)$. The terms are therefore independent of y and the summation with respect to y merely amounts to a multiplication by g. Thus

$$\langle \psi^G, \phi \rangle_G = h^{-1} \sum_{z \in G} \psi(z)\bar{\phi}(z).$$

But ψ vanishes when z does not belong to H. Hence the summation may be restricted to H, and we find that

$$\langle \psi^G, \phi \rangle_G = h^{-1} \sum_{z \in H} \psi(z)\bar{\phi}(z) = \langle \psi, \phi_H \rangle_H,$$

as required.

3.3. The alternating group A_5

This group, which is also the symmetry group of a regular icosahedron [13, pp. 154–6], will here be viewed as the set of the sixty even permutations of five objects. Thus A_5 is a subgroup of index two in S_5, the full symmetric group of degree five. For the standard terminology and elementary facts about permutations the reader is referred to Chapter 4, where further references are given. To simplify the notation we put

$$G = A_5, \qquad S = S_5$$

in this section only. The first problem is to split G into conjugacy classes. In S the conjugacy classes of the even permutations are represented by the cycle patterns

$$C^*: 1, \qquad (12)(34), \qquad (123), \qquad (12345)$$

comprising

$$h^*: 1, \qquad 15, \qquad 20, \qquad 24$$

elements respectively (see (4.10), p. 108 where h_α should be replaced by h_α^*). Now if two elements x and y of G are conjugate in S, it does not follow that they are also conjugate in G. However, let

$$t^{-1}xt = y, \tag{3.16}$$

where t is any permutation, and suppose that x commutes with some odd permutation s, so that

$$s^{-1}xs = x \quad (s \notin G). \tag{3.17}$$

Then in addition to (3.16) we have that

$$(st)^{-1}x(st) = y,$$

and either t or st is even. Hence x and y remain conjugate if (3.17) holds. For example, (12)(34) commutes with the odd permutation (12); hence the conjugacy class of (12)(34) is the same in S as in G. Likewise, (123) commutes with (45) so that this class also remains unchanged. But the situation is different in the case of the element

$$u = (12345), \tag{3.18}$$

as we shall now show. We have already remarked (p. 72) that if $C(u)$ is the conjugacy class and Z_u the centraliser of u in G, then

$$|C(u)| = [G : Z_u]. \tag{3.19}$$

75

Similarly, if the star refers to the group S,

$$|C^*(u)| = [S : Z_u^*].$$

Now $|S| = 120$ and, as mentioned above, $|C^*(u)| = 24$. It follows that $|Z_u^*| = 5$. Since the five elements

$$1, u, u^2, u^3, u^4 \tag{3.20}$$

surely commute with u, we infer that Z_u^* consists precisely of (3.20). As all these elements are even, we deduce that $Z_u = Z_u^*$, whence, by (3.19), we have that $|C(u)| = 12$. Thus the class of u in S 'splits' in G. In fact, the elements u and u^2 are conjugate in S but not in G. For we have that

$$t^{-1}ut = u^2, \tag{3.21}$$

where

$$t = (2354)$$

is an odd permutation. If there were an even permutation w such that

$$w^{-1}uw = u^2,$$

then tw^{-1}, which is odd, would commute with u. This is impossible, as we have just seen. Iterating the relation (3.21) we find that

$$t^{-2}ut^2 = t^{-1}u^2t = u^4. \tag{3.22}$$

Hence u and u^4 are conjugate in G, and so also are u^2 and u^3.

We are now in the position to describe the conjugacy classes of G; for each class we name a representative and give the size of the class:

C_α:	1	(12)(34)	(123)	u	u^2
h_α:	1	15	20	12	12

Henceforth the classes will be enumerated in this order. The above table tells us that G has five irreducible representations, and we proceed to find their characters.

(1) As usual, the trivial character will be denoted by $\chi^{(1)}$.

(2) Since G is a permutation group, Proposition 3.1 furnishes the character

$$\chi^{(2)}: 4, 0, 1, -1, -1.$$

We find that

$$\langle \chi^{(2)}, \chi^{(2)} \rangle_G = \tfrac{1}{60}(16 + 0 + 20 + 12 + 12) = 1,$$

which shows that $\chi^{(2)}$ is a simple character (Proposition 2.1).

(3) The alternating group A_4 on the objects 1, 2, 3, 4 is a subgroup of G. As we have seen (Table 2.5(b), p. 61), A_4 possesses a linear character ϕ with the following specification:

$$C'_\alpha: 1 \quad (12)(34) \quad (123) \quad (123)^2$$

$$h'_\alpha: 1 \qquad 3 \qquad 4 \qquad 4$$

$$\phi: \; 1 \qquad 1 \qquad \omega \qquad \omega^2,$$

where C'_α denotes a class of A_4 and h'_α its size, and where $\omega = \exp(2\pi i/3)$ so that

$$\omega + \omega^2 = -1.$$

We construct the induced character ϕ^G with the aid of (3.13), namely

$$\phi_\alpha^G = \frac{n}{h_\alpha} \sum_w \phi(w) \quad (w \in C_\alpha \cap A_4).$$

In the present case $n = [G : A_4] = 60/12 = 5$. Going through the five classes of G we obtain that

$$\phi_1^G = 5, \; \phi_2^G = \tfrac{5}{15}(3 \times 1) = 1,$$

$$\phi_3^G = \tfrac{5}{20}(4\omega + 4\omega^2) = -1,$$

$$\phi_4^G = 0, \; \phi_5^G = 0,$$

the last two equations arising from the fact that the classes C_4 and C_5 have no element in common with A_4. Since

$$\langle \phi^G, \phi^G \rangle_G = \tfrac{1}{60}(25 + 15 + 20) = 1,$$

the character ϕ^G is simple and we shall label it $\chi^{(3)}$, thus

$$\chi^{(3)}: 5, 1, -1, 0, 0.$$

(4) In order to continue our search for characters we use the cyclic subgroup

$$U: 1, u, u^2, u^3, u^4$$

generated by (3.18). However, our luck of hitting a simple character at the first shot is now going to forsake us. A non-trivial linear character of U is specified by

$$\lambda(u) = \varepsilon,$$

where ε is an arbitrary fifth root of unity, other than unity. We shall, in the first instance, take

$$\varepsilon = \exp(2\pi i/5), \tag{3.23}$$

but ε may be replaced by ε^2 or ε^3 or ε^4, and all we require is the fact that

$$\varepsilon^4 + \varepsilon^3 + \varepsilon^2 + \varepsilon + 1 = 0.$$

We use (3.13) to compute the induced characters λ^G, and we note that $n = [G:U] = 12$. First, we have to determine the intersections $C_\alpha \cap U$ ($\alpha = 1, 2, 3, 4, 5$). Clearly,

$$C_1 \cap U = \{1\}, \quad C_2 \cap U = C_3 \cap U = \varnothing.$$

Since u and u^4 are conjugate in G, we have that

$$C_4 \cap U = \{u, u^4\},$$

and similarly

$$C_5 \cap U = \{u^2, u^3\}.$$

Straightforward calculations now yield the result that

$$\lambda^G: 12, 0, 0, \varepsilon + \varepsilon^4, \varepsilon^2 + \varepsilon^3.$$

Incidentally, we observe that the values of λ^G are real, because

$$\varepsilon + \varepsilon^4 = \varepsilon + \varepsilon^{-1} = 2\cos(2\pi/5),$$

and

$$\varepsilon^2 + \varepsilon^3 = \varepsilon^2 + \varepsilon^{-2} = 2\cos(4\pi/5).$$

Of course, it cannot be expected that a character of such a high degree is simple. In fact, using (3.23), we find that

$$\langle \lambda^G, \lambda^G \rangle_G = \tfrac{1}{60}\left(144 + 12(\varepsilon + \varepsilon^4)^2 + 12(\varepsilon^2 + \varepsilon^3)^2\right) = 3.$$

Thus λ^G is the sum of three simple characters. We enquire whether the Fourier analysis of λ^G involves any of the simple characters we have previously discovered. Easy computations show that

$$\langle \lambda^G, \chi^{(1)} \rangle = 0, \langle \lambda^G, \chi^{(2)} \rangle = 1, \langle \lambda^G, \chi^{(3)} \rangle = 1.$$

Hence $\lambda^G - \chi^{(2)} - \chi^{(3)}$ is a simple character. We shall denote it by $\chi^{(4)}$, thus

$$\chi^{(4)}: 3, -1, 0, -\varepsilon^2 - \varepsilon^3, -\varepsilon - \varepsilon^4.$$

(5) Finally, on replacing ε by ε^2 we obtain the remaining character $\chi^{(5)}$. Other substitutions $\varepsilon \to \varepsilon^s$ would be permissible but would not lead to new results. Hence the complete character table is as in Table 3.1.

$$A_5:$$

C:	1	(12)(34)	(123)	(12345)	$(12345)^2$
h:	1	15	20	12	12
$\chi^{(1)}$	1	1	1	1	1
$\chi^{(2)}$	4	0	1	-1	-1
$\chi^{(3)}$	5	1	-1	0	0
$\chi^{(4)}$	3	-1	0	$-2\cos(4\pi/5)$	$-2\cos(2\pi/5)$
$\chi^{(5)}$	3	-1	0	$-2\cos(2\pi/5)$	$-2\cos(4\pi/5)$

Table 3.1.

A glance at this table reveals the fact that A_5 has no proper normal subgroup. For if N were such a subgroup, then A_5/N would have a non-trivial simple character, which on being lifted to A_5 would be one of the characters included in this table. But a lifted character has the property that $\chi(1) = \chi(y)$ for some y of N, other than 1. This is plainly not the case except for $\chi^{(1)}$, which proves that A_5 is indeed a *simple* group.

3.4. Normal subgroups

The notion of an induced representation and the reciprocity law apply to any subgroup. But more precise information can be obtained when the subgroup is normal. In this section we consider a group G and a normal subgroup H of order h. Thus if $u \in H$ and $t \in G$, then $tut^{-1} \in H$. Suppose that B is a representation of H, so that

$$B(u)B(v) = B(uv) \quad (u, v \in H).$$

Then it is easily verified that, for each fixed t of G, the matrices

$$B_t(u) = B(tut^{-1})$$

form a representation of H, that is

$$B_t(u)B_t(v) = B_t(uv).$$

If ψ is the character of B, then the character of B_t is

$$\psi_t(u) = \psi(tut^{-1}).$$

We say that the representations B and B_t or their characters ψ and ψ_t, are *conjugate* (with respect to G).

Clearly,

$$\langle \psi_t, \psi_t \rangle_H = \langle \psi, \psi \rangle_H,$$

79

since, for fixed t, the elements tut^{-1} run over the group H when u does. In particular, if ψ is a simple character of H, so is each ψ_t.

From now on we shall confine ourselves to a typical simple character ξ of H. The conjugates ξ_t, though simple, need not be distinct. It may happen that, for certain elements t of G,

$$\xi_t(u) = \xi(u) \quad (u \in H).$$

This will surely be the case when t belongs to H, because ξ is a class function on H. More generally, let W be the set of those elements w of G for which

$$\xi(wuw^{-1}) = \xi(u) \quad (u \in H). \tag{3.24}$$

It is easy to check that W is a subgroup of G. We call W the *inertia group* of ξ, and we note that

$$H \leqslant W \leqslant G.$$

Suppose that $\xi_p = \xi_q$, that is

$$\xi(pup^{-1}) = \xi(quq^{-1}) \quad (u \in H).$$

In this identity we may replace u by $q^{-1}uq$, because H is a normal subgroup. Hence

$$\xi(pq^{-1}uqp^{-1}) = \xi(u) \quad (u \in H).$$

This implies that $pq^{-1} \in W$ or, equivalently, that $Wp = Wq$. Thus the number of distinct conjugates of ξ is equal to the number of cosets of W in G. More precisely, if

$$G = \bigcup_{j=1}^{r} Wt_j \quad (t_1 = 1) \tag{3.25}$$

is a coset decomposition of G with respect to W, then

$$\xi_{t_1}(= \xi), \xi_{t_2}, \ldots, \xi_{t_r}$$

is a complete set of distinct conjugates of ξ.

Next, we consider ξ^G. By (3.10),

$$\xi^G(x) = \frac{1}{h} \sum_{y \in G} \xi(yxy^{-1}),$$

which by virtue of (3.25) can be written as

$$\xi^G(x) = \frac{1}{h} \sum_{w \in W} \sum_{j=1}^{r} \xi(wt_jxt_j^{-1}w^{-1}).$$

Restricting this equation to H we have that

$$\xi_H^G(u) = \frac{1}{h} \sum_{w \in W} \sum_{j=1}^{r} \xi(wt_j ut_j^{-1} w^{-1}).$$

Referring to (3.24) we observe that the terms in the sum are independent of w, whence

$$\xi_H^G(u) = [W:H] \sum_{j=1}^{r} \xi_{t_j}(u). \tag{3.26}$$

Since each ξ_{t_j} is a simple character this must be the (unique) Fourier analysis of ξ_H^G, a noteworthy feature being the fact that all coefficients are equal to $[W:H]$.

We now ask the question: in what circumstances is ξ^G a simple character of G? By the reciprocity law

$$\langle \xi^G, \xi^G \rangle_G = \langle \xi_H^G, \xi \rangle_H.$$

When $j \neq 1$, the characters ξ and ξ_{t_j} are orthogonal, while $\xi = \xi_{t_1}$. Hence on taking the inner product of (3.26) with ξ, we find that

$$\langle \xi^G, \xi^G \rangle_G = [W:H].$$

Hence ξ^G is a simple character if and only if $W = H$. Thus we have the following result.

Proposition 3.2. *Let H be a normal subgroup of G and suppose that ξ is a simple character of H. Then ξ^G is a simple character of G if and only if $\xi_t \neq \xi$, whenever $t \notin H$.*

Another problem that arises in this context is the analysis of χ_H (in H), where χ is a typical simple character of G. When χ_H is expressed as a linear combination of the simple characters of H, at least one non-zero term must occur, and we choose the notation in such a way that the coefficient of ξ does not vanish. Thus we suppose that

$$\langle \xi, \chi_H \rangle_H = e > 0. \tag{3.27}$$

By the reciprocity law we have that $\langle \xi^G, \chi \rangle_G = e$. Hence the analysis of ξ^G is of the form

$$\xi^G = e\chi + e'\chi' + e''\chi'' + \dots, \tag{3.28}$$

where $\chi, \chi', \chi'', \dots$ are simple characters of G and $e' \geqslant 0, e'' \geqslant 0, \dots$ On restricting (3.28) to H we obtain that

$$\xi_H^G = e\chi_H + e'\chi_H' + e''\chi_H'' + \dots. \tag{3.29}$$

Suppose now that η is a simple character of H that is not of the form ξ_t. Then (3.26) makes it evident that $\langle \xi_H^G, \eta \rangle_H = 0$. Taking inner products with η on both sides of (3.29) we find that

$$0 = e\langle \chi_H, \eta \rangle_H + e'\langle \chi'_H, \eta \rangle_H + e''\langle \chi''_H, \eta \rangle_H + \ldots$$

Since the inner product of any two characters is always equal to a non-negative integer, each term of this sum must be zero, and in particular

$$\langle \chi_H, \eta \rangle_H = 0,$$

because $e \neq 0$. Hence the Fourier analysis of χ_H does not involve η and must, therefore, be a linear combination of ξ and its conjugates, say

$$\chi_H = \sum_{j=1}^{r} b_j \xi_{t_j}, \tag{3.30}$$

where

$$b_1 = \langle \chi_H, \xi \rangle_H = e.$$

As a matter of fact, all coefficients in (3.30) are equal. For, writing briefly $t = t_j$, we find that

$$\langle \xi_t, \chi_H \rangle_H = \frac{1}{h} \sum_{u \in H} \xi(tut^{-1})\bar{\chi}(u).$$

Replacing the summation variable by $v = tut^{-1}$, we obtain that

$$\langle \xi_t, \chi_H \rangle_H = \frac{1}{h} \sum_{v \in H} \xi(v)\bar{\chi}(t^{-1}vt)$$

$$= \frac{1}{h} \sum_{v} \xi(v)\bar{\chi}(v) = \langle \xi, \chi_H \rangle_H = e.$$

The induced representations of a normal subgroup were studied in depth by A. H. Clifford [4]. His investigations refer to the representations themselves, and not only to the characters, and have therefore wider applications. What we have established can be summarised as follows:

Theorem 3.2 (*Clifford's Theorem*). *Let H be a normal subgroup of G, and let χ be a simple character of G. Then there exists a simple character ξ of H such that*

$$\chi_H = e \sum_{j=1}^{r} \xi_{t_j},$$

where e is a positive integer and the sum involves a complete set of conjugates of ξ.

3.5. Tensor products

On p. 29 we introduced the tensor product of two square matrices and we established the multiplicative property (1.69). Suppose now that $A(x)$ and $B(x)$ are representations of G. Then

$$(A(x) \otimes B(x))(A(y) \otimes B(y)) = A(xy) \otimes B(xy)$$

for all x and y in G. Thus the tensor product of two representations is again a representation, and this gives us a valuable method for constructing new representations. From the definition we immediately see that

$$\text{tr}(P \otimes Q) = (p_{11} + p_{22} + \ldots + p_{mm}) \, \text{tr} \, Q = (\text{tr} \, P)(\text{tr} \, Q).$$

Hence if $\phi(x)$ and $\psi(x)$ are the characters of $A(x)$ and $B(x)$ respectively, it follows that the character of $A(x) \otimes B(x)$ is $\phi(x)\psi(x)$. Incidentally, we recall that the character of the representation

$$\text{diag}(A(x), B(x)) = A(x) \oplus B(x)$$

is equal to $\phi(x) + \psi(x)$. Thus the set of all characters of G admits addition and multiplication though, in general, neither subtraction nor division. In particular, if

$$\chi^{(1)}, \chi^{(2)}, \ldots, \chi^{(k)}$$

is the complete set of simple characters of G, then each product $\chi^{(i)}\chi^{(j)}$ is a character, usually compound. Hence there are equations

$$\chi^{(i)}\chi^{(j)} = \sum_{s=1}^{k} c_{ijs}\chi^{(s)}, \tag{3.31}$$

where the c_{ijs} are k^3 non-negative integers. They are connected with the structure of G in a manner that will not be pursued here. We merely note that by virtue of the orthogonality relations

$$c_{ijt} = g^{-1} \sum_{\alpha=1}^{k} h_\alpha \chi_\alpha^{(i)} \chi_\alpha^{(j)} \bar{\chi}_\alpha^{(t)}.$$

Example. In the character table for A_5 (p. 79) we have that

$$\chi^{(2)}\chi^{(3)} : 20, 0, -1, 0, 0.$$

A straightforward calculation yields the Fourier analysis

$$\chi^{(2)}\chi^{(3)} = \chi^{(2)} + 2\chi^{(3)} + \chi^{(4)} + \chi^{(5)}.$$

83

In order to study the tensor product of representations in greater detail we require the concept of tensor product of two vector spaces, which we are now going to describe: let V and W be vector spaces over the same field K of characteristic zero. We introduce symbols $\mathbf{v} \otimes \mathbf{w}$, where $\mathbf{v}$ and $\mathbf{w}$ range over V and W respectively, and we consider the set of all finite sums

$$\mathbf{T} = \sum_{i,j} \alpha_{ij}(\mathbf{v}_i \otimes \mathbf{w}_j), \qquad (3.32)$$

where $\alpha_{ij} \in K$, $\mathbf{v}_i \in V$, $\mathbf{w}_j \in W$. Addition of two such sums is carried out in the obvious manner, whereby terms associated with the same symbol $\mathbf{v} \otimes \mathbf{w}$ are combined and their coefficients added; for example

$$\{3(\mathbf{v}_1 \otimes \mathbf{w}_1) - 4(\mathbf{v}_2 \otimes \mathbf{w}_2)\} + \{-(\mathbf{v}_1 \otimes \mathbf{w}_1) + 5(\mathbf{v}_3 \otimes \mathbf{w}_3)\}$$
$$= 2(\mathbf{v}_1 \otimes \mathbf{w}_1) - 4(\mathbf{v}_2 \otimes \mathbf{w}_2) + 5(\mathbf{v}_3 \otimes \mathbf{w}_3).$$

Multiplication of (3.32) by an element λ of K is defined by

$$\lambda \mathbf{T} = \sum_{i,j} \lambda \alpha_{ij}(\mathbf{v}_i \otimes \mathbf{w}_j).$$

We now come to the characteristic features of the tensor product: the symbols $\mathbf{v} \otimes \mathbf{w}$ obey the following axioms:

$$\text{(I)} \quad \begin{cases} (\mathbf{v}_1 + \mathbf{v}_2) \otimes \mathbf{w} = (\mathbf{v}_1 \otimes \mathbf{w}) + (\mathbf{v}_2 \otimes \mathbf{w}) \\ \mathbf{v} \otimes (\mathbf{w}_1 + \mathbf{w}_2) = (\mathbf{v} \otimes \mathbf{w}_1) + (\mathbf{v} \otimes \mathbf{w}_2) \end{cases}$$
$$\text{(II)} \quad \alpha \mathbf{v} \otimes \mathbf{w} = \mathbf{v} \otimes \alpha \mathbf{w} = \alpha(\mathbf{v} \otimes \mathbf{w}), \qquad (3.33)$$

where $\mathbf{v}$, $\mathbf{v}_1$, $\mathbf{v}_2 \in V$, $\mathbf{w}$, $\mathbf{w}_1$, $\mathbf{w}_2 \in W$, $\alpha \in K$. The two axioms express the fact that the product $\mathbf{v} \otimes \mathbf{w}$ is bilinear with respect to $\mathbf{v}$ and $\mathbf{w}$. As a consequence we have that

$$\mathbf{0}_V \otimes \mathbf{w} = \mathbf{v} \otimes \mathbf{0}_W = (\mathbf{0}_V \otimes \mathbf{0}_W), \qquad (3.34)$$

where $\mathbf{0}_V$ and $\mathbf{0}_W$ are the zero vectors, and $\mathbf{v}$ and $\mathbf{w}$ are arbitrary vectors of V and W respectively; any one of the expressions (3.34) can be used as the zero element for the sums (3.32).

The imposition of relations upon a system carries with it the risk that its structure might collapse. To prevent this contingency we introduce one more axiom:

(III) $\quad$ If $\mathbf{v}_1$, $\mathbf{v}_2$, ..., $\mathbf{v}_k$ are linearly independent vectors of V and if $\mathbf{w}_1$, $\mathbf{w}_2$, ..., $\mathbf{w}_l$ are linearly independent vectors of W, then the kl symbols

$$\mathbf{v}_r \otimes \mathbf{w}_s \quad (r = 1, 2, \ldots, k; \, s = 1, 2, \ldots, l)$$

are linearly independent, that is if

$$\sum_{r=1}^{k} \sum_{s=1}^{l} \alpha_{rs}(\mathbf{v}_r \otimes \mathbf{w}_s) = (\mathbf{0}_V \otimes \mathbf{0}_W),$$

then $\alpha_{rs} = 0 \; (r = 1, 2, \ldots, k; \, s = 1, 2, \ldots, l)$.

The set of expressions (3.32), subject to the axioms (I), (II) and (III), is called the *tensor product* of V and W and is denoted by

$$V \otimes W.$$

The above discussion gives only an informal account of the concept of a tensor product. For a rigorous treatment the reader should consult a textbook on modern algebra, for example S. MacLane & G. Birkhoff [16, pp. 320ff.]. One of the reasons why a more precise definition is necessary, stems from the obligation to prove that the axioms (I) to (III) are mutually compatible. Without recourse to the abstract theory this point can be settled alternatively by the construction of a concrete model for $V \otimes W$, at least when V and W are finite-dimensional vector spaces over K. Thus suppose that V is the space of all m-tuples over K, written as row vectors,

$$\mathbf{v} = (v_1, v_2, \ldots, v_m)$$

and that W is the vector space of n-tuples over K,

$$\mathbf{w} = (w_1, w_2, \ldots, w_n).$$

We interpret $\mathbf{v} \otimes \mathbf{w}$ as the Kronecker product (see p. 29) of the $1 \times m$ and $1 \times n$ matrices $\mathbf{v}$ and $\mathbf{w}$, thus

$$\begin{aligned}
\mathbf{V} \otimes \mathbf{W} &= (v_1 \mathbf{w}, v_2 \mathbf{w}, \ldots, v_m \mathbf{w}) \\
&= (v_1 w_1, v_1 w_2, \ldots, v_1 w_n, v_2 w_1, \ldots, v_m w_n),
\end{aligned} \tag{3.35}$$

that is, $\mathbf{v} \otimes \mathbf{w}$ is an mn-row vector. Here and elsewhere, when the elements of a set are enumerated by pairs of positive integers we adopt the *lexical ordering*

$$11, 12, \ldots, 1n, \quad 21, 22, \ldots, 2n, \ldots, mn.$$

When $\mathbf{v} \otimes \mathbf{w}$ is defined by (3.35), the axioms (I) and (II) are evidently satisfied. It is less obvious that (III) is also true; we defer its verification to Exercise 9 at the end of this chapter.

Henceforth we shall assume that the tensor product of two vector spaces exists and that it has the properties postulated above.

An immediate consequence of the axioms is the following result, in which the notation (1.11) is used.

Proposition 3.3 (*Basis Theorem for Tensor Products*). *Let*

$$V = [e_1, e_2, \ldots, e_m]$$

and (3.36)

$$W = [f_1, f_2, \ldots, f_n].$$

Then the mn products

$$(e_r \times f_s) \quad (r = 1, 2, \ldots, m; \ s = 1, 2, \ldots, n) \tag{3.37}$$

form a basis for $V \otimes W$. Hence

$$\dim(V \otimes W) = (\dim V)(\dim W).$$

Proof. Each vector v of V is a linear combination of $e_1, \ldots, e_m$ and each vector w of W is a linear combination of $f_1, \ldots, f_n$. The bilinearity of $v \otimes w$ implies that

$$v \otimes w = \sum_{r=1}^{m} \sum_{s=1}^{n} c_{rs}(e_r \times f_s), \tag{3.38}$$

where c_{rs} are suitable elements of K. More generally, (3.38) may be applied to each term of (3.32). Hence an arbitrary element of $V \otimes W$ can be expressed as

$$T = \sum_r \sum_s t_{rs}(e_r \otimes f_s), \tag{3.39}$$

where $t_{rs} \in K$. This shows that the products (3.37) span $V \otimes W$; moreover, they are linearly independent by virtue of (III) and so form a basis.

The formula (3.39) sets up a one-to-one correspondence between the elements of $V \otimes W$ and the set $M_{m,n}(F)$ all $m \times n$ matrices

$$T = (t_{rs})$$

over K. We denote this correspondence by

$$T \leftrightarrow T. \tag{3.40}$$

It is, in fact, an isomorphism between the vector spaces $V \otimes W$ and $M_{m,n}(K)$, because if

$$T_i \leftrightarrow T_i \quad (i = 1, 2)$$

then

$$\lambda_1 T_1 + \lambda_2 T_2 \leftrightarrow \lambda_1 T_1 + \lambda_2 T_2,$$

where λ_1 and λ_2 are arbitrary elements of K.

Of course, the correspondence (3.40) is based on the assumption that the bases mentioned in (3.36) remain fixed. If we change the bases, the same element $\mathbf{T}$ of $V \otimes W$ will be associated with a different matrix, say

$$\mathbf{T} \leftrightarrow \hat{T}. \tag{3.41}$$

It is useful to find an explicit formula for $\hat{T}$. Thus suppose the new bases $(\hat{\mathbf{e}}_\alpha)$ and $(\hat{\mathbf{f}}_\beta)$ are given by

$$\mathbf{e}_r = \sum_{\alpha=1}^{m} p_{r\alpha}\hat{\mathbf{e}}_\alpha \quad (r=1,2,\ldots,m)$$

$$\mathbf{f}_s = \sum_{\beta=1}^{n} q_{s\beta}\hat{\mathbf{f}}_\beta \quad (s=1,2,\ldots,n), \tag{3.42}$$

where

$$P = (p_{r\alpha}) \quad \text{and} \quad Q = (q_{s\beta})$$

are $m \times m$ and $n \times n$ invertible matrices. Substituting in (3.39) and using bilinearity we obtain that

$$\mathbf{T} = \sum_{r,s} \sum_{\alpha,\beta} p_{r\alpha} t_{rs} q_{s\beta}(\hat{\mathbf{e}}_\alpha \otimes \hat{\mathbf{f}}_\beta);$$

more briefly,

$$\mathbf{T} = \sum_{\alpha,\beta} \hat{t}_{\alpha\beta}(\hat{\mathbf{e}}_\alpha \otimes \hat{\mathbf{f}}_\beta),$$

where

$$\hat{T} = P'TQ, \tag{3.43}$$

P' being the transpose of P.

Next, suppose that V and W are G-modules, affording respectively the representations $A(x)$ and $B(x)$, $(x \in G)$, relative to the bases $(\mathbf{e}_r)$ and $(\mathbf{f}_s)$. Thus

$$\mathbf{e}_i x = \sum_{j} a_{ij}(x)\mathbf{e}_j, \quad \mathbf{f}_r x = \sum_{s} b_{rs}(x)\mathbf{f}_s \tag{3.44}$$

$(i,j=1,2,\ldots,m;\ r,s=1,2,\ldots,n)$, and

$$A(x)A(y) = A(xy), \quad B(x)B(y) = B(xy).$$

We turn $V \otimes W$ into a G-module by setting

$$(\mathbf{e}_i \otimes \mathbf{f}_r)x = \mathbf{e}_i x \otimes \mathbf{f}_r x = \sum_{j,s} a_{ij}(x)b_{rs}(x)(\mathbf{e}_j \otimes \mathbf{f}_s). \tag{3.45}$$

The coefficients on the right-hand side form the elements of the matrix

$$A(x) \otimes B(x), \tag{3.46}$$

whose rows and columns are enumerated by the pairs of integers (i, r) and (j, s) respectively. By virtue of (3.31), the matrices (3.46) do indeed form a representation of G. Hence we have the following result.

Proposition 3.4. *Let V be a G-module which affords the representation $A(x)$ relative to the basis $(\mathbf{e}_i)$, and let W be a G-module which affords the representation $B(x)$ relative to the basis $(\mathbf{f}_r)$. Then $V \otimes W$ affords the representation $A(x) \otimes B(x)$ relative to the basis $(\mathbf{e}_i \otimes \mathbf{f}_r)$.*

For the remainder of this section we shall be concerned with the special case in which $V = W$. Thus

$$V \otimes V \tag{3.47}$$

is a G-module which affords the representation

$$A(x) \otimes A(x) \tag{3.48}$$

of degree m^2. A typical element of (3.47) can be written as

$$\mathbf{T} = \sum_{i,j=1}^{m} t_{ij}(\mathbf{e}_i \otimes \mathbf{e}_j) \tag{3.49}$$

and is called a *tensor of rank 2*. Relative to the basis $(\mathbf{e}_i)$ we have the correspondence

$$\mathbf{T} \leftrightarrow T, \tag{3.50}$$

where

$$\mathbf{T} = (t_{ij}).$$

But if the basis is changed to $(\hat{\mathbf{e}}_r)$ by the transformation

$$\mathbf{e}_i = \sum_r p_{ir}\hat{\mathbf{e}}_r,$$

then (3.50) becomes

$$\mathbf{T} \leftrightarrow \hat{T},$$

where

$$\hat{T} = P'TP. \tag{3.51}$$

If the action of G on V is given by

$$\mathbf{e}_i x = \sum_r a_{ir}(x)\mathbf{e}_r,$$

it follows from (3.49) that the action of G on $V \otimes V$ can be described by

$$\mathbf{T}x = \sum_{i,j} \sum_{r,s} t_{ij} a_{ir}(x) a_{js}(x) (\mathbf{e}_r \times \mathbf{e}_s).$$

In terms of the correspondence (3.50) this result can be expressed as

$$\mathbf{T}x \leftrightarrow A'(x) T A(x). \tag{3.52}$$

When $m > 1$, it turns out that the G-module $V \otimes V$ is always reducible; in particular, we shall describe two important submodules:

(i) If T is a symmetric matrix, that is if

$$T' = T, \tag{3.53}$$

we transfer this notion to tensors: thus the tensor $\mathbf{T}$, given in (3.49) is said to be *symmetric* if

$$t_{ij} = t_{ji} \quad (i,j = 1, 2, \ldots, m).$$

This definition makes sense because it is independent of the basis; for if (3.53) holds, then

$$\hat{T}' = \hat{T}$$

and conversely. The collection of all symmetric tensors is denoted by

$$V^{(2)}.$$

Clearly, $V^{(2)}$ is a vector space, because if $\mathbf{T}_1$ and $\mathbf{T}_2$ are symmetric tensors, so is

$$\lambda_1 \mathbf{T}_1 + \lambda_2 \mathbf{T}_2,$$

where $\lambda_1, \lambda_2 \in K$.

Moreover, $V^{(2)}$ is a G-module; for if T is a symmetric matrix, so is the matrix on the right of (3.52). In other words, if

$$\mathbf{T} \in V^{(2)}, \quad \text{then } \mathbf{T}x \in V^{(2)}.$$

It is easy to construct a basis for $V^{(2)}$: let

$$E_{ij} \quad (i,j = 1, 2, \ldots, m)$$

be the matrix that has a unit in the (i,j)th position and zeros elsewhere. Then the matrices

$$E_{ii} \qquad (i = 1, 2, \ldots, m)$$

$$E_{ij} + E_{ji} \quad (1 \leqslant i \leqslant j \leqslant m) \tag{3.54}$$

are symmetric and linearly independent; they span the space of all symmetric matrices, because

$$\sum_{i=1}^{m} t_{ii}E_{ii} + \sum_{i<j} t_{ij}(E_{ij} + E_{ji}) = (t_{ij}),$$

provided that $t_{ij} = t_{ji}$. Hence (3.54) form a basis for the space of symmetric matrices. By virtue of the correspondence (3.50) the symmetric tensors

$$\begin{aligned}
\mathbf{E}_{ii} &= \mathbf{e}_i \otimes \mathbf{e}_i && (i = 1, 2, \ldots, m) \\
\mathbf{E}_{ij} &= (\mathbf{e}_i \otimes \mathbf{e}_j) + (\mathbf{e}_j \otimes \mathbf{e}_i) && (1 \leqslant i < j \leqslant m)
\end{aligned}$$ (3.55)

form a basis of $V^{(2)}$. Since the number of their elements is equal to

$$m + \tfrac{1}{2}m(m-1) = \tfrac{1}{2}m(m+1),$$

it follows that

$$\dim V^{(2)} = \tfrac{1}{2}m(m+1).$$ (3.56)

We arrange the basis elements of $V^{(2)}$ in lexical order, thus

$$\mathbf{E}_{11}, \mathbf{E}_{12}, \ldots, \mathbf{E}_{1m}, \quad \mathbf{E}_{22}, \mathbf{E}_{23}, \ldots, \mathbf{E}_{mm};$$

or more briefly

$$\mathbf{E}_{ij} \quad (1 \leqslant i \leqslant j \leqslant m).$$

Relative to this basis, the G-module $V^{(2)}$ affords a representation of degree $\tfrac{1}{2}m(m+1)$ which we denote by

$$A^{(2)}(x) = (a^{(2)}(i, j; k, l; x)) \quad (x \in G).$$

The entries in this matrix are given by

$$\mathbf{E}_{ij}x = \sum_{k \leqslant l} a^{(2)}(i, j; k, l; x)\mathbf{E}_{kl},$$

and we have that

$$A^{(2)}(x)A^{(2)}(y) = A^{(2)}(xy).$$

The matrix $A^{(2)}(x)$ is called the *second induced* or *second Schläflian matrix* of $A(x)$.

Example 1. Let $m = 2$, and suppose that, for a particular element x of G, we have that

$$A(x) = \begin{pmatrix} a & b \\ c & d \end{pmatrix}.$$

Thus,

$$\mathbf{e}_1 x = a\mathbf{e}_1 + b\mathbf{e}_2, \quad \mathbf{e}_2 x = c\mathbf{e}_1 + d\mathbf{e}_2.$$

Then

$$\mathbf{E}_{11} x = (a\mathbf{e}_1 + b\mathbf{e}_2) \otimes (a\mathbf{e}_1 + b\mathbf{e}_2)$$

$$= a^2 \mathbf{E}_{11} + ab\mathbf{E}_{12} + b^2 \mathbf{E}_{22},$$

$$\mathbf{E}_{12} x = (a\mathbf{e}_1 + b\mathbf{e}_2) \otimes (c\mathbf{e}_1 + d\mathbf{e}_2) + (c\mathbf{e}_1 + d\mathbf{e}_2) \otimes (a\mathbf{e}_1 + b\mathbf{e}_2)$$

$$= 2ac\mathbf{E}_{11} + (ad + bc)\mathbf{E}_{12} + 2bd\mathbf{E}_{22},$$

$$\mathbf{E}_{22} x = (c\mathbf{e}_1 + d\mathbf{e}_2) \otimes (c\mathbf{e}_1 + d\mathbf{e}_2)$$

$$= c^2 \mathbf{E}_{11} + cd\mathbf{E}_{12} + d^2 \mathbf{E}_{22}.$$

Hence

$$\begin{pmatrix} a & b \\ b & d \end{pmatrix}^{(2)} = \begin{pmatrix} a^2 & ab & b^2 \\ 2ac & ad + bc & 2bd \\ c^2 & cd & d^2 \end{pmatrix}.$$

Next, we wish to express the character of $A^{(2)}(x)$ in terms of the character of $A(x)$. We confine ourselves to finite groups G. In that case, for each $x \in G$, the matrix $A(x)$ is periodic and hence diagonalisable (Corollary, p. 57). Thus relative to a suitable basis, which depends on x, we can write

$$A(x) = \text{diag}(\alpha_1, \alpha_2, \ldots, \alpha_m), \tag{3.57}$$

that is

$$\mathbf{e}_i x = \alpha_i \mathbf{e}_i \quad (i = 1, 2, \ldots, m). \tag{3.58}$$

Using the basis (3.55) we see that the action of x on $V^{(2)}$ is described by the equations

$$\mathbf{E}_{ii} x = \alpha_i^2 \mathbf{E}_{ii} \quad (i = 1, 2, \ldots, m),$$

$$\mathbf{E}_{ij} x = \alpha_i \alpha_j \mathbf{E}_{ij} \quad (1 \leqslant i < j \leqslant m).$$

Thus

$$A^{(2)}(x) = \text{diag}(\alpha_1^2, \alpha_1 \alpha_2, \ldots, \alpha_2^2, \alpha_2 \alpha_3, \ldots, \alpha_m^2).$$

Denoting the character of $A^{(2)}(x)$ by $\phi^{(2)}(x)$ we have that

$$\phi^{(2)}(x) = \sum_{i \leqslant j} \alpha_i \alpha_j,$$

or

$$\phi^{(2)}(x) = \tfrac{1}{2}(\alpha_1 + \ldots + \alpha_m)^2 + \tfrac{1}{2}(\alpha_1^2 + \ldots + \alpha_m^2).$$

Since by (3.57)

$$(A(x))^2 = A(x^2) = \text{diag}(\alpha_1^2, \alpha_2^2, \ldots, \alpha_m^2),$$

91

we can write

$$\phi^{(2)}(x) = \tfrac{1}{2}\{(\phi(x))^2 + \phi(x^2)\} \tag{3.59}$$

where

$$\phi(x) = \alpha_1 + \alpha_2 + \ldots + \alpha_m$$

is the character of $A(x)$.

We can now turn our attention to a second submodule of $V \otimes V$. The matrix T is called *anti-symmetric* or *skew-symmetric* if

$$T' = -T.$$

We transfer this term to tensors of rank 2, and we say that the tensor

$$\mathbf{T} = \sum_{i,j} t_{ij}(\mathbf{e}_i \otimes \mathbf{e}_j)$$

is *anti-symmetric* or *skew-symmetric* if

$$t_{ij} = -t_{ji} \quad (i,j = 1, 2, \ldots, m); \tag{3.60}$$

in particular, $t_{ii} = 0$ $(i = 1, 2, \ldots, m)$. The most general anti-symmetric tensor is of the form

$$\mathbf{T} = \sum_{i<j} t_{ij}\{(\mathbf{e}_i \otimes \mathbf{e}_j) - (\mathbf{e}_j \otimes \mathbf{e}_i)\}. \tag{3.61}$$

It is evident from (3.51) that the property of being anti-symmetric does not depend on the choice of basis. The collection of all anti-symmetric tensors is denoted by

$$V^{(1^2)}. \tag{3.62}$$

Clearly this is a vector space; for if $\mathbf{T}_1$ and $\mathbf{T}_2$ are anti-symmetric tensors, so is

$$\lambda_1 \mathbf{T}_1 + \lambda_2 \mathbf{T}_2,$$

where $\lambda_1, \lambda_2 \in K$. We deduce from (3.61) that the tensors

$$\mathbf{D}_{ij} = (\mathbf{e}_i \otimes \mathbf{e}_j) - (\mathbf{e}_j \otimes \mathbf{e}_i) \quad (1 \leqslant i < j \leqslant m) \tag{3.63}$$

form a basis of (3.62). Hence

$$\dim V^{(1^2)} = \tfrac{1}{2}m(m-1). \tag{3.64}$$

Moreover, $V^{(1^2)}$ is a G-module; for if T is an anti-symmetric matrix, so is the matrix on the right-hand side of (3.52). In other words,

$$\text{if } \mathbf{T} \in V^{(1^2)}, \quad \text{then } \mathbf{T}x \in V^{(1^2)}.$$

Relative to the basis (3.63), the G-module $V^{(1^2)}$ affords a representation of degree $\frac{1}{2}m(m-1)$ which we denote by

$$A^{(1^2)}(x) = (a^{(1^2)}(i,j;k,l;x)) \quad (x \in G).$$

The entries in this matrix are given by

$$\mathbf{D}_{ij}x = \sum_{k<l} a^{(1^2)}(i,j;k,l;x)\mathbf{D}_{kl},$$

and we have that

$$A^{(1^2)}(x)A^{(1^2)}(y) = A^{(1^2)}(xy). \tag{3.65}$$

The matrix $A^{(1^2)}(x)$ is the *second compound* of $A(x)$, which was studied in Exercise 8 of Chapter 1, where it was found that

$$a^{(1^2)}(i,j;k,l;x) = a_{ik}(x)a_{jl}(x) - a_{il}(x)a_{jk}(x).$$

The character of $A^{(1^2)}(x)$ is denoted by

$$\phi^{(1^2)}(x) \quad (x \in G).$$

For a fixed element x of G we may assume that (3.57) holds. Then, relative to the basis (3.63), the action of x on $V^{(1^2)}$ is given by

$$\mathbf{D}_{ij}x = \alpha_i\alpha_j\mathbf{D}_{ij} \quad (1 \leqslant i < j \leqslant m).$$

Thus

$$A^{(1^2)}(x) = \operatorname{diag}(\alpha_1\alpha_2, \alpha_1\alpha_3, \ldots, \alpha_{m-1}\alpha_m),$$

whence

$$\phi^{(1^2)}(x) = \sum_{i<j} \alpha_i\alpha_j,$$

or

$$\phi^{(1^2)}(x) = \frac{1}{2}\{(\phi(x))^2 - \phi(x^2)\}. \tag{3.66}$$

Finally, we observe that the two submodules of $V \otimes V$ complement each other. For every square matrix T can be written as the sum of a symmetric and an anti-symmetric matrix; indeed

$$T = \frac{1}{2}(T+T') + \frac{1}{2}(T-T'), \tag{3.67}$$

where

$$\frac{1}{2}(T+T')$$

is symmetric and

$$\frac{1}{2}(T-T')$$

is anti-symmetric. The decomposition (3.68) is unique; for if

$$T = S + R,$$

93

where $S' = S$ and $R' = -R$, it follows that

$$T' = S - R,$$

whence

$$S = \tfrac{1}{2}(T + T'), \quad R = \tfrac{1}{2}(T - T').$$

Also, the zero matrix is the only matrix which is both symmetric and anti-symmetric. Using the notion of a direct sum we may therefore state that

$$V \otimes V = V^{(2)} \otimes V^{(1^2)}.$$

It is convenient to summarise some of the foregoing results in the following:

Theorem 3.3. *Let G be a finite group and suppose that V is a G-module over K of degree m which affords the representation $A(x)$ with character $\phi(x)$. When $m \geqslant 2$, the tensor produce $V \otimes V$ has the following two G-submodules (possibly among others):*

(i) $V^{(2)}$ is of degree $\tfrac{1}{2}m(m+1)$ and consists of all symmetric tensors; it affords a representation $A^{(2)}(x)$ of G with character

$$\phi^{(2)}(x) = \tfrac{1}{2}\{(\phi(x))^2 + \phi(x^2)\},$$

and

(ii) $V^{(1^2)}$ is of degree $\tfrac{1}{2}m(m-1)$ and consists of all anti-symmetric tensors; it affords a representation $A^{(1^2)}(x)$ of G with character

$$\phi^{(1^2)}(x) = \tfrac{1}{2}\{(\phi(x))^2 - \phi(x^2)\}.$$

Moreover,

$$V \otimes V = V^{(2)} \otimes V^{(1^2)}.$$

3.6. Mackey's Theorem

In this section we shall consider the tensor product of two representations that have been induced from representations of arbitrary subgroups. This problem was extensively studied by G. W. Mackey [15]. His investigations deal with the representations themselves. But since, in this book, we confine ourselves to the situation in which complete reducibility is ensured, it suffices for our purposes to study the characters only.

Let L and M be subgroups of G of orders l and m. Suppose that $P(u)$ $(u \in L)$ and $Q(v)$ $(v \in M)$ are representations of L and M respectively, with characters $\phi(u)$ and $\psi(v)$. As usual, we put

$$P(x) = 0, \quad \phi(x) = 0 \quad \text{when } x \notin L,$$

$$Q(x) = 0, \quad \psi(x) = 0 \quad \text{when } x \notin M.$$

We are interested in the representation

$$P^G \otimes Q^G \tag{3.68}$$

of G, where P^G and Q^G are the representations induced from P and Q respectively. As we have seen (p. 83), the character of (3.68) is $\phi^G \psi^G$. Using (3.10) we have that

$$\phi^G(x)\psi^G(x) = (lm)^{-1} \sum_{p,q \in G} \phi(pxp^{-1})\psi(qxq^{-1}). \tag{3.69}$$

Our aim is to recast the right-hand side of (3.69) in such a way that each term appears as the induced character of a subgroup that is related to the problem. Before we can carry out these manipulations we have to recall a few facts about the double coset decomposition

$$G = Lt_1 M \cup Lt_2 M \cup \ldots \cup Lt_r M \tag{3.70}$$

of G with respect to L and M [13, pp. 53–4]. For the moment let us concentrate on a single double coset

$$T = LtM.$$

We associate with T the group

$$D = t^{-1}Lt \cap M$$

and prove the following result.

Lemma. *When u and v run independently over L and M, the lm elements of the form utv sweep out T in such a way that each element of T is covered d times, where $d = |t^{-1}Lt \cap M|$.*

Proof. Suppose that

$$utv = u_1 tv_1, \tag{3.71}$$

where $u, u_1 \in L$ and $v, v_1 \in M$. This happens if and only if

$$t^{-1}u_1^{-1}ut = v_1 v^{-1} = w,$$

where $w \in D$. Thus (3.71) is equivalent to

$$u_1 = utw^{-1}t^{-1}, \qquad v_1 = wv \quad (w \in D),$$

and there are precisely d choices for w. This proves the lemma.

Next, we observe that $P(tzt^{-1})$ is a representation of $t^{-1}Lt$, where a typical element of this group will be denoted by $z = t^{-1}ut$ $(u \in L)$. The tensor product

$$P(twt^{-1}) \otimes Q(w) \quad (w \in D)$$

may therefore be regarded as a representation of D. The character of this representation is

$$\xi(w) = \phi(twt^{-1})\psi(w) \quad (w \in D).$$

With the usual convention that $\xi(x) = 0$ when $x \notin D$, we obtain the induced character

$$\xi^G(x) = d^{-1} \sum_{q \in G} \xi(qxq^{-1}) = d^{-1} \sum_{q \in G} \phi(tqxq^{-1}t^{-1})\psi(qxq^{-1}).$$

Whether y belongs to L or not, it is always true that $\phi(uyu^{-1}) = \phi(y)$, where u is an arbitrary element of L. Hence we may replace t by ut and simultaneously q by vq, where v is a fixed element of M; for if q runs over G, so does vq. However, the latter change is ineffectual for ψ because $\psi(vyv^{-1}) = \psi(y)$ in all cases. Thus

$$\xi^G(x) = d^{-1} \sum_{q \in G} \phi(utvqxq^{-1}v^{-1}t^{-1}u^{-1})\psi(qxq^{-1}).$$

We shall now sum this equation over all u in L and all v in M. This will multiply the left-hand side by lm, while, in accordance with the lemma, the products utv on the right will cover T exactly d times. Hence

$$\xi^G(x) = (lm)^{-1} \sum_{q \in G} \sum_{w \in T} \phi(wqxq^{-1}w^{-1})\psi(qxq^{-1}). \qquad (3.72)$$

We carry out this procedure for each double coset

$$T_\rho = Lt_\rho M \qquad (3.73)$$

in (3.70) and for the corresponding subgroup

$$D_\rho = t_\rho^{-1}Lt_\rho \cap M$$

$(\rho = 1, 2, \ldots, r)$. Thus we construct the character

$$\xi_\rho(w_\rho) = \phi(t_\rho w_\rho t_\rho^{-1})\psi(w_\rho) \quad (w_\rho \in D_\rho)$$

of D_ρ. By analogy with (3.72)

$$\xi_\rho^G(x) = (lm)^{-1} \sum_{q \in G} \sum_{w_\rho \in T_\rho} \phi(w_\rho qxq^{-1}w_\rho^{-1})\psi(qxq^{-1}).$$

Finally, we observe that for any function $f(x)$ $(x \in G)$ we have the identity

96

$$\sum_{\rho=1}^{r} \sum_{w_\rho \in T_\rho} f(w_\rho) = \sum_{y \in G} f(y),$$

which is an immediate consequence of (3.73) and (3.70). Hence

$$\sum_{\rho=1}^{r} \xi_\rho^G(x) = (lm)^{-1} \sum_{q \in G} \sum_{y \in G} \phi(yqxq^{-1}y^{-1})\psi(qxq^{-1}).$$

On changing the summation variable in the inner sum from y to $p = yq$ we obtain that

$$\sum_{\rho=1}^{r} \xi_\rho^G(x) = (lm)^{-1} \sum_{p,q \in G} \phi(pxp^{-1})\psi(qxq^{-1}). \qquad (3.74)$$

Comparison of (3.69) and (3.74) leads to the result which we formulate as follows:

Theorem 3.4. (*G. W. Mackey*). *Let L and M be subgroups of G and let*

$$G = Lt_1M \cup Lt_2M \cup \ldots \cup Lt_rM$$

be the double coset decomposition of G relative to L and M. Define the subgroups

$$D_\rho = t_\rho^{-1} Lt_\rho \cap M \quad (\rho = 1, 2, \ldots, r).$$

Suppose that $\phi(u)$ $(u \in L)$ and $\psi(v)$ $(v \in M)$ are characters of L and M respectively. Then

$$\xi_\rho(w_\rho) = \phi(t_\rho w_\rho t_\rho^{-1})\psi(w_\rho) \quad (w_\rho \in D_\rho)$$

is a character of D_ρ, and the induced characters ϕ^G, ψ^G and ξ_ρ^G satisfy the relations

$$\phi^G(x)\psi^G(x) = \sum_{\rho=1}^{r} \xi_\rho^G(x) \quad (x \in G).$$

Exercises

1. The group G of order 21 is defined by the relations

$$a^7 = b^3 = 1, \qquad b^{-1}ab = a^2.$$

Verify that the conjugacy classes are

$$C_1 = (1), \qquad C_2 = (a, a^2, a^4), \qquad C_3 = (a^3, a^5, a^6),$$

$$C_4 = (a^\alpha b), \qquad C_5 = (a^\alpha b^2) \quad (\alpha = 0, 1, 2, \ldots, 6).$$

Prove that the character table is as in Table 3.2.

	C_1	C_2	C_3	C_4	C_5
h_α	1	3	3	7	7
$\chi^{(1)}$	1	1	1	1	1
$\chi^{(2)}$	1	1	1	ω	ω^2
$\chi^{(3)}$	1	1	1	ω^2	ω
$\chi^{(4)}$	3	η	$\bar\eta$	0	0
$\chi^{(5)}$	3	$\bar\eta$	η	0	0

where $\omega^2+\omega+1=0$ and $\eta^2+\eta+2=0$.

Table 3.2.

2. The group G of order p^3, where p is a prime, is defined by the relations

$$a^{p^2}=b^p=1, \qquad b^{-1}ab=a^{1+p}.$$

Show that $Z=\mathrm{gp}\{a^p\}$ is the centre of G. Prove that G has p^2 linear characters $\lambda^{(r,s)}$ $(r,s=0,1,\ldots,p-1)$ and $p-1$ characters $\chi^{(t)}$ $(t=1,2,\ldots,p-1)$ of degree p. If $\varepsilon=\exp(2\pi i/p)$, the values of the characters are as follows:

$$\lambda^{(r,s)}(a^\alpha b^\beta Z)=\varepsilon^{\alpha r+\beta s} \qquad (\alpha,\beta=0,1,\ldots,p-1)$$

$$\begin{cases}\chi^{(t)}(a^{np})=p\varepsilon^{tn} & (n=0,1,\ldots,p-1)\\ \chi^{(t)}(x)=0 & \text{if } x\notin Z.\end{cases}$$

3. The group of order p^3, where p is a prime, is defined by the relations

$$a^p=b^p=c^p=1, \qquad ab=ba, \qquad ac=ca, \qquad c^{-1}bc=ab.$$

Prove that G has p^2 linear characters $\lambda^{(r,s)}$ $(r,s=0,1,\ldots,p-1)$ and $p-1$ characters $\chi^{(t)}$ $(t=1,2,\ldots,p-1)$ of degree p. If $\varepsilon=\exp(2\pi i/p)$, the values of the characters are as follows:

$$\lambda^{(r,s)}(b^\beta c^\gamma Z)=\varepsilon^{\beta r+\gamma s} \qquad (\beta,\gamma=0,1,\ldots,p-1)$$

$$\begin{cases}\chi^{(t)}(a^\alpha)=p\varepsilon^{t\alpha} & (\alpha=0,1,\ldots,p-1)\\ \chi^{(t)}(x)=0 & \text{if } x\notin Z.\end{cases}$$

Thus this group has the same character table as the group studied in the preceding exercise.

4. Let L and H be subgroups of G such that $L<H<G$, and let ϕ be a character of L. Put $\psi=\phi^H$. Prove that $\phi^G=\psi^G$. Thus the process of induction may be carried out in stages (*transitivity of induction*).

5. Let ϕ be a character of a finite group G, and let ψ be a character of a subgroup H of G. Prove that

$$\{\phi_H\psi\}^G=\phi\psi^G.$$

6. [**6**, p. 68]. Suppose that the finite group G contains an Abelian subgroup A. Prove that if $f^{(i)}$ $(i = 1, 2, \ldots, k)$ are the degrees of the simple characters of G, then

$$f^{(i)} \leq [G : A].$$

7. Let $W = G \times H$ be the direct product of two finite groups G and H. Thus the elements of W are the pairs (x, y), where $x \in G$ and $y \in H$, and multiplication in W is defined by $(x, y)(x', y') = (xx', yy')$. Let $A(x)$ and $B(y)$ be representations of G and H respectively. Prove that the *Kronecker product* $C((x, y)) = A(x) \otimes B(y)$ is a representation of W. Prove that if A and B are absolutely irreducible, so is C.

8. [**6**, p. 70]. If $A(x)$ is a representation of G with character $\phi(x)$, then the rth *Kronecker power* $A^{(r)}(x) = A(x) \otimes A(x) \otimes \ldots \otimes A(x)$ (r factors) $(r = 0, 1, \ldots)$ has character $\{\phi(x)\}^r$, where $A^{(0)}(x)$ is to be identified with the trivial representation.

Suppose now that $A(x)$ is a faithful representation of degree m and that $\phi(x)$ takes t distinct values

$$\omega_1 (= m), \omega_2, \ldots, \omega_t$$

as x ranges over G. Define

$$U_s = \{x \in G \mid \phi(x) = \omega_s\} \quad (s = 1, 2, \ldots, t).$$

Corresponding to an arbitrary simple character χ put

$$z_s = g^{-1} \sum_{x \in U_s} \bar{\chi}(x) \quad (s = 1, 2, \ldots, t)$$

and

$$e_r = \langle \chi, \phi^r \rangle \quad (r = 0, 1, \ldots, t-1).$$

Show that

$$e_r = \sum_{s=1}^{t} \omega_s^r z_s$$

and prove that there exists at least one value of r such that $e_r \neq 0$. Thus, up to equivalence, each irreducible representation of G occurs in at least one of the Kronecker powers $A^{(r)}(x)$ $(r = 0, 1, \ldots, t-1)$.

9. Let V and W be the spaces of row vectors of dimensions m and n respectively over the field K. Suppose that

$$V = [\mathbf{e}_1, \mathbf{e}_2, \ldots, \mathbf{e}_m],$$

where

$$\mathbf{e}_1 = (1, 0, \ldots, 0), \quad \mathbf{e}_2 = (0, 1, \ldots, 0), \ldots, \quad \mathbf{e}_m = (0, 0, \ldots, 1)$$

is the standard basis of dimension m, and that

$$\mathbf{W} = [\mathbf{f}_1, \mathbf{f}_2, \ldots, \mathbf{f}_n],$$

where

$$\mathbf{f}_1 = (1, 0, \ldots, 0), \quad \mathbf{f}_2 = (0, 1, \ldots, 0), \ldots, \quad \mathbf{f}_n = (0, 0, \ldots, 1)$$

is the standard basis of dimension n.

(i) Show that the mn row vectors $\mathbf{e}_i \otimes \mathbf{f}_j$ $(i = 1, 2, \ldots, m; j = 1, 2, \ldots, n)$, when regarded as elements of $V \otimes W$, are linearly independent over K.

(ii) Suppose that $\mathbf{v}_1, \mathbf{v}_2, \ldots, \mathbf{v}_k$ are linearly independent vectors of V and that $\mathbf{w}_1, \mathbf{w}_2, \ldots, \mathbf{w}_l$ are linearly independent vectors of W. Prove that $\mathbf{v}_r \otimes \mathbf{w}_s$ $(r = 1, 2, \ldots, k; s = 1, 2, \ldots, l)$ are linearly independent vectors of $V \otimes W$.

10. Let $A = (a_{ij})$ be an $m \times n$ matrix over a field K. The *second induced* or *Schläflian* matrix $A^{(2)}$ of A is defined as follows: suppose that

$$V = [\mathbf{e}_1, \ldots, \mathbf{e}_m] \quad \text{and} \quad W = [\mathbf{f}_1, \ldots, \mathbf{f}_n]$$

are vector spaces of dimensions m and n respectively over K. Associate with A the linear map $\alpha: V \to W$ specified by

$$\mathbf{e}_i \alpha = \sum_j a_{ij} \mathbf{f}_j.$$

This map induces a map $\alpha^{(2)}: V^{(2)} \to W^{(2)}$ given by

$$\mathbf{E}_{ir} \alpha^{(2)} = \sum_j \sum_s a^{(2)}(i, j; r, s) \mathbf{F}_{js}$$

where

$$\mathbf{E}_{ii} = \mathbf{e}_i \otimes \mathbf{e}_i \quad (1 \leqslant i \leqslant m),$$

$$\mathbf{E}_{ir} = (\mathbf{e}_i \otimes \mathbf{e}_r) + (\mathbf{e}_r \otimes \mathbf{e}_i) \quad (1 \leqslant i < r \leqslant m),$$

$$\mathbf{F}_{jj} = \mathbf{f}_j \otimes \mathbf{f}_j \quad (1 \leqslant j \leqslant n),$$

$$\mathbf{F}_{js} = (\mathbf{f}_j \otimes \mathbf{f}_s) + (\mathbf{f}_s \otimes \mathbf{f}_j) \quad (1 \leqslant j < s \leqslant n),$$

Thus the matrix

$$A^{(2)} = (a^{(2)}(i, j; r, s))$$

has $\frac{1}{2}m(m+1)$ rows and $\frac{1}{2}n(n+1)$ columns. Let $B = (b_{kl})$ be an $n \times t$ matrix. Prove that

$$(AB)^{(2)} = A^{(2)} B^{(2)}.$$

11. Referring to the character table of the group A_5 on p. 79 show that

(i) $$(\chi^{(2)})^{(2)} = \chi^{(1)} + \chi^{(2)} + \chi^{(3)}$$

(ii) $$(\chi^{(2)})^{(1^2)} = \chi^{(4)} + \chi^{(5)}$$

(iii) $$(\chi^{(4)})^{(1^2)} = \chi^{(4)}.$$

PERMUTATION GROUPS

4.1. Transitive groups

We continue the discussion of permutation groups, which was briefly begun in § 1.5 (p. 17). Let G be a permutation group of degree n, acting on the symbols

$$1, 2, \ldots, n. \tag{4.1}$$

Thus a typical element of G is a permutation

$$x : \alpha \to \alpha x \quad (\alpha = 1, 2, \ldots, n),$$

where $1x, 2x, \ldots, nx$ are the symbols (4.1) in some order.

Definition 4.1. *The group G is said to be transitive on $1, 2, \ldots, n$ if, for every pair (α, β), where $1 \leqslant \alpha, \beta \leqslant n$, there exists at least one permutation $x_{\alpha\beta}$ in G with the property that*

$$\alpha x_{\alpha\beta} = \beta.$$

We note that α and β need not be distinct and that $x_{\alpha\beta}$ is not required to be unique.

Transitivity is ensured by what, at first sight, appears to be a weaker condition.

Criterion. *The group G is transitive if and only if for each $\alpha (1 \leqslant \alpha \leqslant n)$ there exists at least one permutation p_α in G with the property that*

$$1 p_\alpha = \alpha. \tag{4.2}$$

The condition of the criterion is certainly necessary, as it corresponds to the pair $(1, \alpha)$ mentioned in the definition. Conversely, if p_α satisfies (4.2), then $\alpha(p_\alpha^{-1} p_\beta) = \beta$, and the group is transitive.

Example. Consider the following subgroups of S_4:

$$A : 1, \quad (12)(34), \quad (13)(24), \quad (14)(23),$$

and

$$B : 1, \quad (12), \quad (34), \quad (12)(34).$$

The criterion tells us that A is transitive. But B is clearly not transitive, as it contains no permutation which maps 1 into 3. Incidentally, we have that $A \cong B$ as abstract groups.

Definition 4.2. *Those permutations of G which leave the symbol 1 fixed form a subgroup of G, which is called the stabiliser of 1 (in G).*

We have singled out the symbol 1 for convenience. For a transitive group, this does not amount to a serious loss of generality, as is shown in the following proposition.

Proposition 4.1. *Let G be a transitive group on $1, 2, \ldots, n$ and suppose that $1p_\alpha = \alpha$ ($\alpha = 1, 2, \ldots, n$). If H is the stabiliser of 1, then the stabiliser of α is the group $p_\alpha^{-1} H p_\alpha$, which is isomorphic to H.*

Proof. Let H and H_α be the stabilisers of the symbols 1 and α respectively. First, let u be an arbitrary element of H, that is $1u = 1$. Then

$$\alpha(p_\alpha^{-1} u p_\alpha) = \alpha,$$

which implies that $p_\alpha^{-1} H p_\alpha \subset H_\alpha$. Next, if $v \in H_\alpha$, that is, $\alpha v = \alpha$, we have that

$$1(p_\alpha v p_\alpha^{-1}) = 1.$$

Hence $p_\alpha H_\alpha p_\alpha^{-1} \subset H$, that is $H_\alpha \subset p_\alpha^{-1} H p_\alpha$. This proves that $H_\alpha = p_\alpha^{-1} H p_\alpha$, as required. Henceforth we shall assume that G is transitive and that the permutations p_α ($\alpha = 1, 2, \ldots, n$) have been chosen once and for all. Then the permutation x transforms 1 into α if and only if x lies in the coset $H p_\alpha$. As α varies from 1 to n we obtain all the permutations of G. Thus we have the coset decomposition

$$G = H p_1 \cup H p_2 \cup \ldots \cup H p_n.$$

It follows that

$$[G : H] = n,$$

and we note that the degree of a transitive group divides the order of the group.

Every permutation group has a natural representation in terms of permutation matrices (p. 18). Denoting this representation by $N(x)$ and its trace by $\nu(x)$ we have seen ((1.48)) that

$$\nu(x) = \textit{number of symbols fixed by } x.$$

As before, denote the stabiliser of 1 by H. Let $\xi^{(1)}$ be the trivial character of H. We assert that

$$\xi^{(1)G}(x) = \sum_{\alpha=1}^{n} \xi^{(1)}(p_\alpha x p_\alpha^{-1}) = \nu(x). \qquad (4.3)$$

For each term in the sum is either zero or unity, and it is equal to unity if and only if $p_\alpha x p_\alpha^{-1} \in H$, that is $x \in p_\alpha^{-1} H p_\alpha = H_\alpha$, which means that x leaves α fixed. Thus the sum counts the number of fixed symbols, which proves (4.3).

It is obvious that, if $\chi^{(1)}$ is the trivial character of G, then its restriction to H is the trivial character of H, that is

$$\chi_H^{(1)} = \xi^{(1)}.$$

This equality may be expressed differently as

$$\langle \chi_H^{(1)}, \xi^{(1)} \rangle_H = 1.$$

On applying the Reciprocity Theorem (p. 74) to the last equation and using (4.3) we deduce that

$$\langle \chi^{(1)}, \xi^{(1)G} \rangle_G = \langle \chi^{(1)}, \nu \rangle_G = 1. \qquad (4.4)$$

It is worth-while writing out the inner product explicitly. The result then reads as follows:

Proposition 4.2. *Let G be a transitive permutation group of order g. If $\nu(x)$ is the natural character of G, then*

$$\sum_{x \in G} \nu(x) = g.$$

It is noteworthy that this formula does not involve the degree of G. The Fourier analysis of ν is of the form

$$\nu = \chi^{(1)} + e_2 \chi^{(2)} + e_3 \chi^{(3)} + \ldots, \qquad (4.5)$$

the first coefficient on the right being equal to unity by virtue of (4.4). If we wish to obtain information about the other coefficients, we have to impose more stringent conditions on G.

Definition 4.3. *A permutation group G, acting on the symbols $1, 2, \ldots, n$ is said to be doubly transitive if corresponding to any two pairs of symbols (α, β) and (γ, δ), where $\alpha \neq \beta$ and $\gamma \neq \delta$, there exists at least one permutation x in G satisfying*

$$\alpha x = \gamma, \qquad \beta x = \delta.$$

Suppose now that G is doubly transitive and let H be the stabiliser of 1 in G. A typical permutation of H has the form

$$u = \begin{pmatrix} 1 & 2 & 3 & \ldots & n \\ 1 & \alpha_2 & \alpha_3 & \ldots & \alpha_n \end{pmatrix}.$$

We associate with u the permutation

$$u^1 = \begin{pmatrix} 2 & 3 & \ldots & n \\ \alpha_2 & \alpha_3 & \ldots & \alpha_n \end{pmatrix}.$$

As u ranges over H, then u^1 ranges over a group H^1 acting on $2, 3, \ldots, n$. Clearly, the correspondence

$$u \leftrightarrow u^1 \qquad (4.6)$$

establishes an isomorphism between H and H^1. It is easy to see that H^1 is transitive. For if β and γ are any symbols, other than 1, we know that there exists a permutation v in G with the properties that

$$1v = 1, \qquad \beta v = \gamma.$$

Hence v belongs to H and, using the correspondence (4.6), we have that $\beta v^1 = \gamma$. Thus H^1 is transitive. Evidently,

$$\nu(u) = 1 + \nu^1(u^1),$$

where ν^1 is the natural character of H^1. Summing over u or, equivalently, over u^1, we obtain that

$$\sum_{u \in H} \nu(u) = h + \sum_{u^1 \in H^1} \nu^1(u^1),$$

where

$$h = |H| = |H^1|.$$

Applying Proposition 4.2 to H^1 we infer that the sum on the right is equal to h. Thus

$$\sum_{u \in H} \nu(u) = 2h,$$

that is

$$\langle \nu_H, \xi^{(1)} \rangle_H = 2.$$

Invoking the Reciprocity Theorem once again and using (4.3) we deduce that

$$\langle \nu, \nu \rangle_G = 2.$$

It follows that the coefficients in (4.5) satisfy

$$2 = 1 + e_2^2 + e_3^2 + \dots.$$

This means that precisely one of the coefficients $e_2, e_3, \dots$ is equal to unity, all the others being zero. Hence $\nu - \chi^{(1)}$ is a simple character. We record this conclusion:

Theorem 4.1. *If G is a doubly transitive group with natural character $\nu(x)$, then the function*

$$\nu(x) - 1$$

is a simple character of G.

Example. As an illustration we construct the character table of the symmetric group S_4. The conjugacy classes are represented by

$$1, \quad (12), \quad (123), \quad (12)(34), \quad (1234),$$

comprising

$$1, \quad 6, \quad 8, \quad 3, \quad 6$$

elements respectively [**13**, p. 138]. Hence S_4 possesses five irreducible representations. As usual, we denote the trivial character by $\chi^{(1)}$. The alternating character is

$$\chi^{(2)}: 1, -1, 1, 1, -1,$$

because the elements (12) and (1234) are odd while 1, (123) and (12)(34) are even. Next, we exploit the fact that S_4 has the normal subgroup [**13**, p. 138]

$$V: 1, (12)(34), (13)(24), (14)(23).$$

With a convenient choice of coset representatives we may assume that

$$S_4/V: V, V(12), V(13), V(23), V(123), V(132).$$

Thus

$$S_4/V \cong S_3.$$

Hence any one of the characters of S_3 may be lifted so as to yield a character of S_4 (§ 2.5). The only character that gives a useful result is the character of degree two in Table 2.2 (p. 50). In the notation of Theorem 2.6, we have that

$$\chi_0^{(3)}(V) = 2, \quad \chi_0^{(3)}(V(12)) = 0, \quad \chi_0^{(3)}(V(123)) = -1,$$

which gives the following values of the lifted character:

$$\chi^{(3)}(1) = \chi^{(3)}((12)(34)) = 2, \qquad \chi^{(3)}((12)) = 0, \qquad \chi^{(3)}((123)) = -1.$$

Also, since

$$(1234) = (12)(34)(13) \in V(13),$$

we have that

$$\chi^{(3)}((1234)) = \chi_0^{(3)}(V(13)) = \chi_0^{(3)}(V(12)) = 0.$$

Hence we have obtained the character

$$\chi^{(3)}: 2, 0, -1, 2, 0.$$

Evidently, S_4 is doubly transitive. We may therefore apply Theorem 4.1. This gives us the character

$$\chi^{(4)}: 3, 1, 0, -1, -1.$$

Finally, we note that $\chi^{(2)}\chi^{(4)} = \chi^{(5)}$ is a simple character distinct from those constructed so far, namely

$$\chi^{(5)}: 3, -1, 0, -1, 1.$$

The complete character table is given in Table 4.1.

$$S_4:$$

C_α	1	(12)	(123)	(12)(34)	(1234)
h_α	1	6	8	3	6
$\chi^{(1)}$	1	1	1	1	1
$\chi^{(2)}$	1	-1	1	1	-1
$\chi^{(3)}$	2	0	-1	2	0
$\chi^{(4)}$	3	1	0	-1	-1
$\chi^{(5)}$	3	-1	0	-1	1

Table 4.1.

4.2. The symmetric group

The somewhat haphazard methods we have developed so far are insufficient to obtain all the simple characters of the symmetric group S_n, when n is an arbitrary positive integer. It is one of the most remarkable achievements of Frobenius that, only four years after the discovery of group characters in 1896, he succeeded in obtaining all the characters of S_n, at least in principle. More precisely, he constructed a set of generating functions whose coefficients reveal the full character table. It must, however, be admitted that the expansion of the generating functions tends to be cumbersome, except in special circumstances. A few years later, Issai Schur developed an alternative version of the character theory for the symmetric group. He constructed a set of generating functions which, in a sense, are dual to those of Frobenius. These Schur functions have received a great deal of attention [12], as they play an important part in the more advanced study of the symmetric group. We shall devote the remainder of this chapter to giving an account of Frobenius's work in a slightly modified form.

We recall that each permutation of S_n can be resolved into a product of disjoint cycles in a unique manner save for the order of the cycle factors. A cycle involving a single symbol indicates that this symbol remains fixed. Two elements of S_n are conjugate if and only if they have the same cycle pattern [13, p. 131]. For example, in S_6, the permutations

$$x = (146)(35)(2) \quad \text{and} \quad y = (243)(16)(5)$$

have the same cycle pattern and are therefore conjugate; indeed, $t^{-1}xt = y$, where

$$t = \begin{pmatrix} 1 & 2 & 3 & 4 & 5 & 6 \\ 2 & 5 & 1 & 4 & 6 & 3 \end{pmatrix}.$$

Thus each conjugacy class C_α of S_n is determined by its cycle pattern comprising, say, α_1 cycles of degree 1, α_2 cycles of degree 2, and so on. Accordingly, the *specification* of C_α will be described by the formula

$$\alpha : \alpha_1 + 2\alpha_2 + 3\alpha_3 + \ldots + n\alpha_n = n, \tag{4.7}$$

where $\alpha_1, \alpha_2, \ldots, \alpha_n$ are non-negative integers.

Alternatively, each permutation of C_α is the product of, say, u cycles whose degrees, in some order, are $p_1, p_2, \ldots p_u$ respectively. Hence C_α is determined by the *partition*

$$p : p_1 + p_2 + \ldots + p_u = n,$$

where the parts are positive integers. For formal reasons it is often convenient to add $n - u$ zero terms and to define a partition of n as a sum of n non-negative integers, thus

$$p : p_1 + p_2 + \ldots + p_n = n, \tag{4.8}$$

which is abbreviated to

$$|p| = n.$$

107

No distinction is made between partitions that differ merely by the arrangement of their terms. Hence in order to achieve uniqueness we may impose the conditions

$$p_1 \geqslant p_2 \geqslant \ldots \geqslant p_n \geqslant 0. \tag{4.9}$$

Let k be the number of conjugacy classes of S_n. There is no simple expression for k as a function of n, but in view of the foregoing discussion we know that k is the number of solutions of (4.7), or else that k is the number of solutions of (4.8) and (4.9).

When C_α is specified by (4.7), then

$$|C_\alpha| = h_\alpha,$$

where h_α is given by Cauchy's formula [13, p. 132], namely

$$h_\alpha = \frac{n!}{1^{\alpha_1}\alpha_1! 2^{\alpha_2}\alpha_2! \ldots n^{\alpha_n}\alpha_n!}. \tag{4.10}$$

In S_n each element is conjugate to its inverse, since x and x^{-1} possess the same cycle pattern. Hence if ϕ is a character of S_n we have that

$$\phi(x) = \phi(x^{-1}) = \bar{\phi}(x).$$

It follows that all characters of S_n have real values. In fact, it will transpire that these values are (rational) integers (p. 123).

4.3. Induced characters of S_n

Our ultimate aim is to compute the simple characters

$$\chi^{(1)}, \chi^{(2)}, \ldots, \chi^{(k)} \tag{4.11}$$

of S_n. As a first step we shall equip ourselves with a large stock of compound characters by constructing induced characters of suitable subgroups. Corresponding to an arbitrary partition

$$p: p_1 + p_2 + \ldots + p_n = n$$

we form the subgroup

$$H_p = S_{p_1} \times S_{p_2} \times \ldots \times S_{p_n}.$$

Here S_{p_1} acts on the symbols $1, 2, \ldots, p_1$; S_{p_2} acts on the symbols $p_1 + 1, \ldots, p_1 + p_2$; S_{p_3} acts on the symbols $p_1 + p_2 + 1, \ldots, p_1 + p_2 + p_3$, and so on, with the understanding that S_{p_j} is the unit group when $p_j = 0$. Evidently

$$|H_p| = p_1! p_2! \ldots p_n!. \tag{4.12}$$

Let ε be the trivial character of H_p, and let $\phi^{(p)}$ be the character of S_n induced by ε. By (3.13) its value on C_α is given by

$$\phi_\alpha^{(p)} = (1/h_\alpha)[S_n : H_p] \sum_z \varepsilon(z),$$

where z ranges over $C_\alpha \cap H_p$. Hence we have that

$$\phi_\alpha^{(p)} = (1/h_\alpha)[S_n : H_p]|C_\alpha \cap H_p|. \tag{4.13}$$

Only the last factor on the right requires further attention. Let

$$u = u_1 u_2 \ldots u_n \tag{4.14}$$

be a typical element of H_p where $u_j \in S_{p_j}$ $(j = 1, 2, \ldots, n)$, and suppose that u_j is the product of α_{j1}1-cycles, α_{j2}2-cycles, α_{j3}3-cycles, ... We shall determine the conditions on the integers α_{ij} which ensure that u belongs to C_α. First, the total number of i-cycles must add up to α_i. Thus

$$A : \alpha_i = \alpha_{1i} + \alpha_{2i} + \ldots + \alpha_{ni} \quad (i = 1, 2, \ldots, n). \tag{4.15}$$

Secondly, we stipulate that u_j involves p_j symbols, that is

$$B : p_j = \alpha_{j1} + 2\alpha_{j2} + \ldots + n\alpha_{jn} \quad (j = 1, 2, \ldots, n). \tag{4.16}$$

In the equations A and B, the numbers α_i and p_j are prescribed, and the matrix (α_{ji}) is unknown.

Next, we ask how many elements of $H_p \cap C_\alpha$ correspond to a particular solution (α_{ji}). For a fixed value of j, the equation (4.16) specifies a conjugacy class of S_{p_j}. By (4.10), this class contains

$$h^{(j)}(\alpha_{ji}) = p_j! / (1^{\alpha_{j1}} \alpha_{j1}! 2^{\alpha_{j2}} \alpha_{j2}! \ldots n^{\alpha_{jn}} \alpha_{jn}!)$$

elements. Letting j range from 1 to n we obtain

$$\prod_{j=1}^{n} h^{(j)}(\alpha_{ji})$$

elements of $C_\alpha \cap H_p$, all corresponding to the same matrix (α_{ji}). Hence

$$|C_\alpha \cap H_p| = \sum_{A,B} \prod_{j=1}^{n} h^{(j)}(\alpha_{ji}),$$

where the summation is extended over all solutions of A and B. Substituting this expression in (4.13) we find, after a short calculation, that

$$\phi_\alpha^{(p)} = \sum_{A,B} \prod_{j=1}^{n} \frac{\alpha_j!}{\alpha_{j1}! \alpha_{j2}! \ldots \alpha_{jn}!}. \tag{4.17}$$

Thus for each partition we have constructed a compound character of S_n.

109

It is significant that the number of these compound characters is equal to the number of simple characters. This justifies the hope that ultimately we shall succeed in isolating the simple characters from the collection of compound characters. However, a rather formidable apparatus is required for this purpose. Some of the necessary tools will be furnished in the next sections.

4.4. Generalised characters

If $\chi^{(1)}, \chi^{(2)}, \ldots, \chi^{(k)}$ is the complete set of simple characters of a group G, then an arbitrary character of G is a function of the form

$$\phi = e_1 \chi^{(1)} + e_2 \chi^{(2)} + \ldots + e_k \chi^{(k)},$$

where $e_1, e_2, \ldots, e_k$ are non-negative integers. It is convenient to extend this notion as follows:

Definition 4.4. *A class function on G of the form*

$$\xi = u_1 \chi^{(1)} + u_2 \chi^{(2)} + \ldots + u_k \chi^{(k)}, \tag{4.18}$$

where $u_1, u_2, \ldots, u_k$ are integers (positive, negative or zero) is called a generalised character of G.

When negative integers occur among the coefficients in (4.18), then there is no matrix representation of G whose trace is equal to ξ. But the formula for the inner product remains valid for generalised characters. Thus if

$$\eta = v_1 \chi^{(1)} + v_2 \chi^{(2)} + \ldots + v_k \chi^{(k)},$$

the orthogonality relations for the simple characters imply that

$$\langle \xi, \eta \rangle = u_1 v_1 + u_2 v_2 + \ldots + u_k v_k.$$

In particular

$$\langle \xi, \xi \rangle = u_1^2 + u_2^2 + \ldots + u_k^2.$$

Hence, if $\langle \xi, \xi \rangle = 1$, then exactly one of the coefficients in (4.18) is equal to ± 1, while the other coefficients are equal to zero. Thus we have the following

Proposition 4.3. *Let ξ be a generalised character such that $\langle \xi, \xi \rangle = 1$. Then $\xi = \pm \chi$, where χ is a simple character. If, in addition, $\xi(1) > 0$, then $\xi = \chi$.*

More generally, if

$$\xi^{(1)}, \xi^{(2)}, \ldots, \xi^{(s)} \tag{4.19}$$

are s generalised characters such that

$$\langle \xi^{(i)}, \xi^{(j)} \rangle = \delta_{ij}, \quad \xi^{(i)}(1) > 0$$

$(i, j = 1, 2, \ldots, s)$, then (4.19) consists of s distinct simple characters. In particular, if $s = k$, all simple characters are comprised in (4.19).

4.5. Skew-symmetric polynomials

Let

$$x = (x_1, x_2, \ldots, x_n)$$

be n indeterminates. If

$$p = (p_1, p_2, \ldots, p_n) \qquad (4.20)$$

is an n-tuple of non-negative integers, we put

$$(x|p) = x_1^{p_1} x_2^{p_2} \ldots x_n^{p_n}. \qquad (4.21)$$

A polynomial in x may then be written as

$$P(x) = \sum_p a_p (x|p), \qquad (4.22)$$

involving a finite number of terms, the coefficients a_p being real or complex numbers. We recall that two polynomials are regarded as equal if and only if they have the same coefficients. The polynomial P is said to be homogeneous of degree m if $|p| = m$ for each p, that is if p is a partition of m.

Corresponding to any permutation

$$\sigma : i \to i\sigma \quad (i = 1, 2, \ldots, n)$$

we define

$$x\sigma = (x_{1\sigma}, x_{2\sigma}, \ldots, x_{n\sigma})$$

and

$$p\sigma = (p_{1\sigma}, p_{2\sigma}, \ldots, p_{n\sigma}).$$

Since the factors in (4.21) may be rearranged arbitrarily, it follows that

$$(x\sigma|p\sigma) = (x|p). \qquad (4.23)$$

We say that P is a *symmetric polynomial* if

$$P(x\sigma) = P(x)$$

111

for all $\sigma \in S_n$, and P is called *skew-symmetric* (*anti-symmetric, alternating*) if

$$P(x\sigma) = [\sigma]P(x), \tag{4.24}$$

where $[\sigma]$ is the alternating character of S_n [**13**, p. 134], that is

$$[\sigma] = \begin{cases} 1 & \text{if } \sigma \text{ is even} \\ -1 & \text{if } \sigma \text{ is odd.} \end{cases}$$

From now on we shall assume that $n \geqslant 2$.

Replacing x by $x\sigma^{-1}$ we can write (4.24) as

$$P(x) = [\sigma]P(x\sigma^{-1}). \tag{4.25}$$

Formally, it may be assumed that the summation in (4.22) is extended over all n-tuples of non-negative integers, with the proviso that only a finite number of the coefficients a_p is non-zero. Substituting (4.22) in (4.25) we obtain that

$$\sum_p a_p(x|p) = [\sigma]\sum_p a_p(x\sigma^{-1}|p) = [\sigma]\sum_p a_p(x|p\sigma),$$

by virtue of (4.23). On comparing the coefficients of $(x|p\sigma)$ we find that

$$a_{p\sigma} = [\sigma]a_p \quad (\sigma \in S_n). \tag{4.26}$$

It follows that a_p is zero, unless the components of p are distinct. For suppose that $p_i = p_j$ ($i \neq j$) and let $\tau = (i\,j)$. Then evidently $p\tau = p$. Since $[\tau] = -1$, we deduce from (4.26) that $a_p = 0$. Hence in a skew-symmetric polynomial the summation in (4.22) may be restricted to n-tuples p of distinct non-negative integers. Suppose that

$$l_1 > l_2 > \ldots > l_n \geqslant 0$$

are the parts of p arranged in decreasing order of magnitude. Then p may be written as

$$p = l\sigma,$$

where

$$l = (l_1, l_2, \ldots, l_n)$$

is a *strictly decreasing* (*s.d.*) *partition* of n and σ is a suitable permutation. Thus an arbitrary skew-symmetric polynomial can be expressed in the form

$$P = \sum_l \sum_\sigma a_{l\sigma}(x|l\sigma),$$

where l runs over all s.d. partitions and σ ranges over S_n. Using (4.26) we obtain that

$$P = \sum_l a_l \sum_\sigma [\sigma](x \mid l\sigma).$$

We recall the definition of a determinant from first principles, namely

$$\det(c_{ij}) = \sum_\sigma [\sigma] c_{1\sigma,1} c_{2\sigma,2} \ldots c_{n\sigma,n}.$$

Hence, with the same coefficients as in (4.22),

$$P = \sum_l a_l V_l, \tag{4.27}$$

where

$$V_l = \sum_\sigma [\sigma](x \mid l\sigma) = \det(x_j^{l_i}),$$

or explicitly

$$V_l = \begin{vmatrix} x_1^{l_1} & x_2^{l_1} & \ldots & x_n^{l_1} \\ x_1^{l_2} & x_2^{l_2} & \ldots & x_n^{l_2} \\ \ldots & \ldots & & \ldots \\ x_1^{l_n} & x_2^{l_n} & \ldots & x_n^{l_n} \end{vmatrix}. \tag{4.28}$$

Thus the determinants (4.28), which are called *alternants*, span the set of all skew-symmetric polynomials. This result may be strengthened as follows:

Proposition 4.4. *The set*

$$V = \{V_l\},$$

where l ranges over all strictly decreasing partitions of n, forms a basis of the skew-symmetric polynomials in n indeterminates.

Proof. In view of (4.27) it only remains to show that

$$\sum_l b_l V_l = 0 \tag{4.29}$$

implies that $b_l = 0$ for all l. For this purpose we use the lexical ordering of n-tuples. We say that $p = (p_1, p_2, \ldots, p_n)$ precedes $q = (q_1, q_2, \ldots, q_n)$ if

113

the first non-zero difference in the sequence

$$p_1 - q_1, p_2 - q_2, \ldots, p_n - q_n$$

is positive. In this case we write $p > q$, and correspondingly we shall say that $(x|p)$ precedes $(x|q)$; similarly for V_p and V_q.

Let $l = (l_1, l_2, \ldots, l_n)$ be a s.d. n-tuple, that is

$$l_1 > l_2 > \ldots > l_n,$$

and let σ be any element of S_n. Then

$$l > l\sigma \qquad (4.30)$$

unless σ is the identity permutation. For $l_1 > l_{1\sigma}$, except when $1\sigma = 1$; when this happens, $l_1 = l_{1\sigma}$ and $l_2 > l_{2\sigma}$, except when $2\sigma = 2$, and so on.

Suppose now that, if possible, there is a relation (4.29), in which not all coefficients are zero. Without loss of generality, we may drop zero terms in (4.29), if any, and we may assume that V_l is the highest term in the lexical ordering which has a non-zero coefficient. Then, by virtue of (4.30), $(x|l)$ is the highest term in the expansion of V_l. But $(x|l)$ also precedes each term in the expansion of V_m, where $l > m$. Hence, when the left-hand side of (4.29) is expanded as a polynomial in $x_1, x_2, \ldots, x_n$, the term $b_l(x|l)$ cannot be cancelled and we arrive at a contradiction. This proves the proposition.

4.6. Generating functions

We introduce the sums of powers

$$s_r = x_1^r + x_2^r + \ldots + x_n^r \quad (r = 1, 2, \ldots). \qquad (4.31)$$

With a given class C_α, specified by

$$\alpha : \alpha_1 + 2\alpha_2 + \ldots + n\alpha_n = n,$$

we associate the product

$$\mathbf{s}_\alpha = s_1^{\alpha_1} s_2^{\alpha_2} \ldots s_n^{\alpha_n}, \qquad (4.32)$$

which we are going to expand as a polynomial in $x_1, x_2, \ldots, x_n$. By the multinomial theorem

$$s_r^{\alpha_r} = \sum \binom{\alpha_r}{\alpha_{1r}, \ldots, \alpha_{nr}} x_1^{r\alpha_{1r}} \ldots x_n^{r\alpha_{nr}},$$

where

$$\binom{\alpha_r}{\alpha_{1r}, \ldots, \alpha_{nr}} = \frac{\alpha_r!}{\alpha_{1r}! \ldots \alpha_{nr}!}$$

and the summation is extended over all non-negative integers $\alpha_{1r}, \ldots, \alpha_{nr}$ satisfying

$$\alpha_{1r} + \ldots + \alpha_{nr} = \alpha_r.$$

As r varies from 1 to n these equations are the conditions A described in (4.15). Substituting in (4.32) we obtain that

$$\mathbf{s}_\alpha = \prod_{r=1}^{n} \sum_A \binom{\alpha_r}{\alpha_{1r} \ldots \alpha_{nr}} x_1^{r\alpha_{r1}} \ldots x_n^{r\alpha_{rn}}.$$

We must now collect those terms which involve a particular product

$$(x|p) = x_1^{p_1} \ldots x_n^{p_n},$$

that is we gather those terms whose exponents satisfy

$$B : p_j = \alpha_{j1} + 2\alpha_{j2} + \ldots + n\alpha_{nj} \quad (j = 1, 2, \ldots, n)$$

(see (4.16)). Thus the coefficient of $(x|p)$ in $\mathbf{s}_\alpha$ is given by

$$\sum_{A,B} \prod_{r=1}^{n} \binom{\alpha_r}{\alpha_{1r} \ldots \alpha_{nr}} = \sum_{A,B} \prod_{r=1}^{n} \binom{\alpha_r}{\alpha_{r1} \ldots \alpha_{rn}},$$

which is identical with the compound character $\phi_\alpha^{(p)}$ found in (4.17). Hence we have shown that

$$\mathbf{s}_\alpha = \sum_{|p|=n} \phi_\alpha^{(p)}(x|p),$$

and we say that $\mathbf{s}_\alpha$ is a *generating function* for these compound characters.

Remarkable though this formula is, it still falls short of our requirements, for what we seek is a generating function for the simple characters. Now if R is any polynomial in $x_1, x_2, \ldots, x_n$ with integral coefficients then

$$\mathbf{s}_\alpha R = \sum_q \xi_\alpha^{(q)}(x|q), \tag{4.33}$$

where the coefficients $\xi_\alpha^{(q)}$ are linear integral combinations of the $\phi_\alpha^{(p)}$ and are therefore generalised characters in the sense of § 4.4. Our problem would be solved if we could choose R in such a way that the coefficients in (4.33) satisfy the orthogonality relations set out in Proposition 4.3.

With a stroke of genius that is surely unsurpassed even among the great masters of algebraic formalism, Frobenius declares, near the beginning of his memoir [9], that he is going to choose for R the *difference product* or *Vandermonde determinant*

$$\Delta = \prod_{i<j} (x_i - x_j) = \begin{vmatrix} x_1^{n-1} & x_2^{n-1} & \cdots & x_n^{n-1} \\ x_1^{n-2} & x_2^{n-2} & \cdots & x_n^{n-2} \\ \cdots & \cdots & & \cdots \\ x_1 & x_2 & \cdots & x_n \\ 1 & 1 & \cdots & 1 \end{vmatrix} = V_e, \qquad (4.34)$$

where e is the partition

$$e: n-1, n-2, \ldots, 1, 0; \qquad |e| = \tfrac{1}{2}n(n-1).$$

Since $\mathbf{s}_\alpha$ is a symmetric polynomial and Δ is skew-symmetric, the product $\mathbf{s}_\alpha \Delta$ is skew-symmetric. Anticipating the reader's curiosity about the motivation for his *tour de force* Frobenius merely offers the laconic remark (adapted from the German [9], p. 519): 'How it occurred to me to turn the symmetric function [that is $\mathbf{s}_\alpha$] into a skew-symmetric one through multiplication by the difference product will emerge clearly from the subsequent proof.'

The skew-symmetric polynomial $\mathbf{s}_\alpha \Delta$ is of degree

$$N = n + \tfrac{1}{2}n(n-1) = \tfrac{1}{2}n(n+1)$$

in the indeterminates. Applying (4.27) we have an expansion of the form

$$\mathbf{s}_\alpha \Delta = \sum_{|l|=N} \xi_\alpha^{(l)} V_l. \qquad (4.35)$$

The orthogonal properties of the coefficients will be established in the next section.

4.7. Orthogonality

In order to establish the orthogonality relations for the functions $\xi^{(l)}$ Frobenius introduced two sets of indeterminates

$$x_1, x_2, \ldots, x_n \quad \text{and} \quad y_1, y_2, \ldots, y_n.$$

Henceforth it will be necessary to indicate which indeterminates are involved in each instance. Thus we put

$$\Delta(x) = \prod_{i<j} (x_i - x_j),$$

$$s_r(x) = x_1^r + x_2^r + \ldots + x_n^r \quad (r = 1, 2, \ldots),$$

$$V_l(x) = \det(x_j^{l_i}),$$

and similarly for $\Delta(y)$, $s_r(y)$ and $V_l(y)$. We shall find it convenient to employ a further single indeterminate t.

In this section we operate with formal infinite power series in these $2n + 1$ indeterminates. Two power series are equal if and only if they have the same coefficients. Addition and multiplication of power series are carried out in the usual way. Identities like

$$\exp(\log P) = P$$

remain valid since they can be proved by comparing coefficients. Questions of convergence do not arise, since we shall never have occasion to substitute numerical values for the indeterminates in an infinite power series.

Consider the expansion

$$\prod_{i,j=1}^{n} (1 - tx_iy_j)^{-1} = \exp\left\{ -\sum_{i,j} \log(1 - tx_iy_j) \right\}$$

$$= \exp\left\{ t \sum_{i,j} x_iy_j + \tfrac{1}{2}t^2 \sum_{i,j} x_i^2y_j^2 + \tfrac{1}{3}t^3 \sum_{i,j} x_i^3y_j^3 + \ldots \right\}$$

$$= \exp\{ts_1(x)s_1(y) + \tfrac{1}{2}t^2s_2(x)s_2(y) + \ldots\}$$

$$= \prod_{r=1}^{\infty} \exp\left\{ \frac{t^r}{r} s_r(x)s_r(y) \right\}$$

$$= \prod_{r=1}^{\infty} \sum_{\alpha_r=0}^{\infty} (s_r^{\alpha_r}(x)s_r^{\alpha_r}(y)t^{r\alpha_r}/r^{\alpha_r}\alpha_r!). \qquad (4.36)$$

Collecting terms in t^u we write

$$\prod_{i,j=1}^{n} (1 - tx_iy_j)^{-1} = \sum_{u=0}^{\infty} G_u(x, y)t^u, \qquad (4.37)$$

and our next task is to determine the precise form of G_u. For this purpose it is convenient to introduce 'restricted' infinite sequences

$$\alpha = (\alpha_1, \alpha_2, \alpha_3, \ldots)$$

of non-negative integers, in which all but a finite number of terms are zero. It is then legitimate to define

$$\mathbf{s}_\alpha(x) = s_1^{\alpha_1}(x)s_2^{\alpha_2}(x)s_3^{\alpha_3}(x)\ldots$$

and similarly for $\mathbf{s}_\alpha(y)$. Also, we put

$$g(\alpha) = (1^{\alpha_1}\alpha_1!2^{\alpha_2}\alpha_2!\ldots)^{-1}, \qquad (4.38)$$

and we use the notation

$$\|\alpha\| = u, \qquad (4.39)$$

where u is a non-negative integer, to express that

$$\alpha_1 + 2\alpha_2 + 3\alpha_3 + \ldots = u.$$

This condition ensures that $\alpha_i = 0$ when $i > u$.

Returning to (4.36) we find that

$$G_u(x, y) = \sum_{\|\alpha\|=u} g(\alpha)\mathbf{s}_\alpha(x)\mathbf{s}_\alpha(y). \qquad (4.40)$$

In order to establish a link with (4.36) we multiply this equation by $\Delta(x)\Delta(y)$; that is, we consider the expression

$$\Delta(x)\Delta(y) \prod_{i,j=1}^{n} (1 - tx_iy_j)^{-1}. \qquad (4.41)$$

It is a remarkable fact that, long before the discovery of group characters, Cauchy had studied a determinant whose value is closely related to (4.41), namely

$$C_0 = \det\left(\frac{1}{1 - x_iy_j}\right) = \Delta(x)\Delta(y) \prod_{i,j=1}^{n} (1 - x_iy_j)^{-1}$$

$(i, j = 1, 2, \ldots, n)$. This formula will be proved in the Appendix, p. 187. Replace x_i and y_j by $t^{\frac{1}{2}}x_i$ and $t^{\frac{1}{2}}y_j$ respectively. Since $\Delta(x)$ and $\Delta(y)$ are of degree $\frac{1}{2}n(n-1)$, we obtain that

$$t^{\frac{1}{2}n(n-1)}\Delta(x)\Delta(y) \prod_{i,j=1}^{n} (1 - tx_iy_j)^{-1} = \det\left(\frac{1}{1 - tx_iy_j}\right) = C, \qquad (4.42)$$

say. Expanding the determinant from first principles we have that

$$C = \sum_{\sigma \in S_n} [\sigma] \prod_{i=1}^{n} (1 - tx_iy_{i\sigma})^{-1}$$

$$= \sum_{\sigma \in S_n} [\sigma] \prod_{i=1}^{n} (1 + tx_iy_{i\sigma} + t^2x_i^2y_{i\sigma}^2 + \ldots). \qquad (4.43)$$

Let C be arranged as a power series in t, say

$$C = \sum_{v=0}^{\infty} H_v(x, y) t^v. \tag{4.44}$$

A typical term of H_v is a product of the form $(x|p)(y\sigma|p)$, where p is a partition of v. Since C is skew-symmetric in x (and in y), only partitions with distinct parts will occur. Hence we may put

$$p = l\rho,$$

where

$$l_1 > l_2 > \ldots > l_n \geqslant 0, \quad |l| = v$$

and ρ is a suitable permutation of S_n. Inspection of (4.43) shows that all such l and ρ appear. Thus

$$H_v = \sum_{\sigma,\rho} \sum_{|l|=v} [\sigma](x|l\rho)(y\sigma|l\rho).$$

By (4.23), we have that $(y\sigma|l\rho) = (y|l\rho\sigma^{-1})$. Put $\pi = \rho\sigma^{-1}$ and sum over π instead of σ. Then $\sigma = \pi^{-1}\rho$, $[\sigma] = [\pi^{-1}][\rho] = [\pi][\rho]$, and

$$H_v = \sum_{|l|=v} \sum_{\pi,\rho} [\pi][\rho](x|l\rho)(y|l\pi),$$

that is

$$H_v = \sum_{|l|=v} V_l(x) V_l(y). \tag{4.45}$$

Combining (4.41) with (4.37) and (4.45) we obtain the identity

$$t^{\frac{1}{2}n(n-1)} \Delta(x) \Delta(y) \sum_{u=0}^{\infty} G_u t^u = \sum_{v=0}^{\infty} H_v t^v.$$

We are only interested in the coefficient of t^N, where $N = \frac{1}{2}n(n+1)$. Thus on the left we take the term in which u is equal to n. Using (4.40) and (4.45) we find that

$$\sum_{\|\alpha\|=n} g(\alpha) s_\alpha(x) \Delta(x) s_\alpha(y) \Delta(y) = \sum_{|l|=N} V_l(x) V_l(y). \tag{4.46}$$

Finally, we write (4.35) for each set of indeterminates, thus

$$s_\alpha(x) \Delta(x) = \sum_{|l|=N} \xi_\alpha^{(l)} V_l(x), \quad s_\alpha(y) \Delta(y) = \sum_{|m|=N} \xi_\alpha^{(m)} V_m(y).$$

119

Substituting in (4.46) and comparing coefficients in the products $V_l(x) V_m(y)$ we deduce that

$$\sum_{\|\alpha\|=n} g(\alpha) \xi_\alpha^{(l)} \xi_\alpha^{(m)} = \delta_{lm}. \tag{4.47}$$

A glance at (4.38) and (4.10) shows that

$$n! g(\alpha) = h_\alpha \quad \text{if } \|\alpha\| = n. \tag{4.48}$$

Hence (4.47) may be written as

$$\langle \xi^{(l)}, \xi^{(m)} \rangle = \delta_{lm}.$$

It follows from Proposition 4.3 that either $\xi^{(l)}$ is a simple character, or else $-\xi^{(l)}$ is a simple character. In order to rule out the second alternative we shall show that $\xi_1^{(l)} > 0$, where $C_1 = \{1\}$. This will be done in the next section, in which we shall derive an explicit formula for $\xi_1^{(l)}$ due to Frobenius.

4.8. The degree formula

Let

$$f^{(l)} = \xi_1^{(l)}$$

be the value of $\xi^{(l)}$ for the class C_1. The cycle pattern of C_1 is given by

$$\alpha_1 = n, \quad \alpha_2 = \ldots = \alpha_n = 0,$$

and the generating formula (4.35) reduces to

$$s_1^n \Delta = \sum_{|l|=N} f^{(l)} V_l.$$

Now

$$s_1^n = \sum_{|r|=n} \frac{n!}{r_1! r_2! \ldots r_n!} (x|r),$$

where r ranges over all (non-negative) partitions of n.

Next, we have that

$$\Delta = V_e = \sum_\sigma [\sigma](x|e\sigma),$$

where

$$e = (n-1, n-2, \ldots, 1, 0).$$

If p and q are any non-negative n-tuples, we have that

$$(x|p)(x|q) = (x|p+q),$$

$p+q$ being the vector sum of p and q. Hence

$$s_1^n \Delta = \sum_r \sum_\sigma n! \left\{ \prod_i r_i! \right\}^{-1} [\sigma](x|e\sigma + r). \tag{4.49}$$

Let l be a s.d. partition of N. The terms which occur in V_l are of the form $(x|l\rho)$, where ρ is an arbitrary permutation. Thus in order to pick out the coefficient of V_l in (4.49) we must impose on r the condition

$$e\sigma + r = l\rho.$$

It is tempting to achieve this by putting

$$r = l\rho - e\sigma. \tag{4.50}$$

But this might introduce negative components into r. However, the difficulty is overcome by adopting the convention that

$$\frac{1}{(-m)!} = 0 \quad \text{when } m > 0.$$

Then the spurious terms that appear in (4.49) are zero and the relation (4.50) may be assumed to hold throughout. On substituting (4.50) in (4.49) we obtain that

$$s_1^n \Delta = \sum_l \sum_{\sigma,\rho} n! \left\{ \prod_i (l_{i\rho} - e_{i\sigma})! \right\}^{-1} [\sigma](x|l\rho).$$

Let the factors of the product be subjected to the permutation ρ^{-1}. It then becomes

$$\prod_i (l_i - e_{i\sigma\rho^{-1}})!$$

Finally, we put $\pi = \sigma\rho^{-1}$ and sum over π instead of σ, noting that $[\sigma] = [\pi][\rho]$. The summation with respect to ρ then yields V_l and

$$s_1^n \Delta = \sum_l \left(\sum_\pi [\pi] n! \left\{ \prod_i (l_i - e_{i\pi})! \right\}^{-1} \right) V_l.$$

Fortunately, the untidy-looking coefficient of V_l, which is $f^{(l)}$, may be more neatly expressed as a determinant from first principles, namely

$$f^{(l)} = n! \det\left(\frac{1}{(l_i - e_j)!} \right),$$

121

or, on multiplying the ith row by $l_i!$ and dividing the determinant by $l_1! l_2! \ldots l_n!$,

$$f^{(l)} = \frac{n!}{l_1! l_2! \ldots l_n!} \det\left(\frac{l_i!}{(l_i - e_j)!}\right). \tag{4.51}$$

It will be shown in the Appendix (p. 187) that the determinant in (4.51) has the value

$$\Delta(l) = \prod_{i<j} (l_i - l_j), \tag{4.52}$$

which is a positive integer because l is strictly decreasing.

This completes the proof of the celebrated degree formula of Frobenius, namely

$$f^{(l)} = \frac{n! \Delta(l)}{l_1! l_2! \ldots l_n!}. \tag{4.53}$$

The expression on the right does not make it obvious that $f^{(l)}$ is an integer, a fact which is known to us for other reasons. An abbreviated version of (4.53) is mentioned in Exercise 3 (p. 138). We have now established that a simple character of S_n is associated with each s.d. partition of $N\left(= \frac{1}{2}n(n+1)\right)$, say

$$l: l_1 > l_2 > \ldots > l_n \geqslant 0; \quad |l| = N. \tag{4.54}$$

These partitions of N can be placed in a one-to-one correspondence with the (weakly) decreasing partitions of n. For if

$$p: p_1 \geqslant p_2 \geqslant \ldots \geqslant p_n \geqslant 0; \quad |p| = n \tag{4.55}$$

is given, we can define

$$l_i = p_i + n - i \quad (i = 1, 2, \ldots, n),$$

whence (4.54) will be satisfied. Conversely, starting with l we put $p_i = l_i - n + i$, which satisfies (4.55).

As we pointed out on p. 108 the number of partitions p is equal to the number of simple characters of S_n. Hence the generating functions (4.35) reveal the complete set of simple characters. The different characters may be labelled either by l or by p, the latter being more commonly used. Accordingly, we shall change our notation and write

$$V^{(p)} = V_l, \qquad \chi^{(p)} = \xi^{(l)}.$$

The main results of Frobenius's theory may be summarised as follows:

Theorem 4.2. *Let* $x_1, x_2, \ldots, x_n$ *be indeterminates and put*

$$s_r = x_1^r + x_2^r + \ldots + x_n^r \quad (r = 1, 2, \ldots), \qquad \Delta = \prod_{i<j} (x_i - x_j).$$

For each partition

$$p: p_1 \geqslant p_2 \geqslant \ldots \geqslant p_n \geqslant 0; \quad |p| = n$$

let

$$V^{(p)} = \det(x_j^{p_i + n - i}).$$

If the conjugacy class C_α *is specified by the cycle pattern*

$$\alpha_1 + 2\alpha_2 + \ldots + n\alpha_n = n,$$

define

$$\mathbf{s}_\alpha = s_1^{\alpha_1} s_2^{\alpha_2} \ldots s_n^{\alpha_n}.$$

Then the values of all simple characters of S_n *appear as coefficients in the generating functions*

$$\mathbf{s}_\alpha \Delta = \sum_p \chi_\alpha^{(p)} V^{(p)}. \tag{4.56}$$

Moreover,

$$f^{(p)} = \chi_1^{(p)} = \frac{n! \prod_{i<j} (p_i - p_j + j - i)}{\prod_i (p_i + n - i)!}. \tag{4.57}$$

It is clear that when the left-hand side of (4.56) is expanded as a skew-symmetric polynomial in the indeterminates, then all coefficients will be integers. As was shown in (4.27), the same coefficients are involved when the polynomial is expressed as a linear combination of the determinants $V^{(p)}$. Hence we have the

Corollary. *All characters of the symmetric group are integral-valued.*

4.9. Schur functions

Since both $V^{(p)}$ and Δ are skew-symmetric polynomials, the quotient

$$F^{(p)} = V^{(p)}/\Delta$$

123

is a symmetric function of the indeterminates. It is in fact a polynomial, because $V^{(p)}$ vanishes when $x_i = x_j$ $(i \neq j)$. Hence $V^{(p)}$ is divisible by each difference and therefore by Δ.

Definition 4.5. *The polynomial*

$$F^{(p)} = \det(x_j^{p_i+n-i})/\det(x_j^{n-i}) \tag{4.58}$$

is called the Schur function, or S-function, corresponding to the partition

$$p: p_1 \geqslant p_2 \geqslant \ldots \geqslant p_n \geqslant 0; \quad |p| = n.$$

These functions were introduced by I. Schur in his treatment of the symmetric group. We can now write (4.56) as

$$\mathbf{s}_\alpha = \sum_p \chi_\alpha^{(p)} F^{(p)}. \tag{4.59}$$

Since the characters are real, the orthogonality relations of the first kind (p. 39) state that

$$\frac{1}{n!} \sum_\alpha h_\alpha \chi_\alpha^{(p)} \chi_\alpha^{(q)} = \delta_{pq}.$$

Hence we deduce from (4.59) that

$$F^{(q)} = \frac{1}{n!} \sum_\alpha h_\alpha \chi_\alpha^{(q)} \mathbf{s}_\alpha, \tag{4.60}$$

which is Schur's formula. Thus while the generating functions of Frobenius furnish the values of all the characters for a particular class, a Schur function is associated with a particular character and displays its values for all classes. It is in this sense that the two theories may be described as dual to each other.

The problem of determining the characters of S_n has now been reduced to the algebraic task of expanding the Schur functions in terms of $s_1, s_2, \ldots$. Although this is a straightforward programme, its execution often involves heavy calculations. Some short-cuts are available if we take advantage of the wealth of information on symmetric functions, which was assembled in the nineteenth century. The following digression serves this purpose.

The Fundamental Theorem on symmetric functions states that every symmetric polynomial in the indeterminates $x_1, x_2, \ldots, x_n$ can be uniquely expressed in terms of the *elementary symmetric* functions

$$\mathscr{C}: c_1, c_2, \ldots, c_n,$$

124

where

$$c_r = \sum_{(\lambda)} x_{\lambda_1} x_{\lambda_2} \dots x_{\lambda_r},$$

the summation being extended over all sets of r integers $(\lambda) = (\lambda_1, \lambda_2, \dots, \lambda_r)$ such that

$$1 \leqslant \lambda_1 < \lambda_2 < \dots < \lambda_r \leqslant n.$$

Two alternative basic sets of symmetric functions are important. They are the *sums of powers*

$$\mathscr{S}: s_1, s_2, s_3, \dots$$

and the *complete symmetric* (or *Wronski*) functions

$$\mathscr{W}: w_1, w_2, w_3, \dots,$$

where

$$w_r = \sum_{(\mu)} x_{\mu_1} x_{\mu_2} \dots x_{\mu_r}, \qquad 1 \leqslant \mu_1 \leqslant \mu_2 \leqslant \dots \leqslant \mu_r \leqslant n.$$

If uniqueness of expansions were desired, it would be necessary to limit $\mathscr{S}$ and $\mathscr{W}$ to the first n members. But we prefer not to impose this restriction. Indeed, it is convenient to extend the above sequences of functions by defining

$$c_0 = 1, c_m = 0 \quad \text{if } m > n \quad \text{or } m < 0$$

and

$$w_0 = 1, \ w_m = 0 \quad \text{if } m < 0.$$

In order to discover the relations between $\mathscr{C}$, $\mathscr{S}$ and $\mathscr{W}$ we introduce the generating function

$$f(t) = \prod_{i=1}^{n} (1 - x_i t) = 1 - c_1 t + c_2 t^2 + \dots + (-1)^n c_n t^n. \qquad (4.61)$$

Using formal power series we find that

$$\frac{1}{f(t)} = \prod_{i=1}^{n} (1 - x_i t)^{-1} = 1 + w_1 t + w_2 t^2 + \dots. \qquad (4.62)$$

On multiplying (4.61) by (4.62) and picking out the coefficient of t^r we obtain that

$$w_r - c_1 w_{r-1} + c_2 w_{r-2} - \dots + (-1)^n c_n w_{r-n} = \begin{cases} 1 & \text{if } r = 0 \\ 0 & \text{if } r \neq 0 \end{cases} \qquad (4.63)$$

Note that this result holds even when r is negative.

125

Next,

$$\log f(-t) = \sum_{i=1}^{n} \log(1+tx_i) = s_1 t - \tfrac{1}{2}s_2 t^2 + \tfrac{1}{3}s_3 t^3 + \dots,$$

whence

$$f(-t) = \prod_{r=1}^{\infty} \exp\left\{(-1)^{r-1}\frac{s_r}{r}t^r\right\} = \prod_{r=1}^{\infty}\left(\sum_{\alpha_r=0}^{\infty}\frac{1}{\alpha_r!}\left\{(-1)^{r-1}\frac{s_r}{r}t^r\right\}^{\alpha_r}\right).$$

Employing the notations (4.38) and (4.39) we deduce after a short calculation that

$$c_m = \sum_{\|\alpha\|=m} g(\alpha)(-1)^{\alpha_2+\alpha_4+\cdots}s_\alpha. \tag{4.64}$$

Again,

$$\log\frac{1}{f(t)} = \sum_{i=1}^{n} -\log(1-tx_i) = s_1 t + \tfrac{1}{2}s_2 t^2 + \tfrac{1}{3}s_3 t^3 + \dots,$$

whence by taking exponentials and comparing coefficients of t^m,

$$w_m = \sum_{\|\alpha\|=m} g(\alpha)s_\alpha. \tag{4.65}$$

Observe the striking similarity between the formulae (4.64) and (4.65): each is turned into the other by changing the signs of $s_2, s_4, \dots$. In particular, when $m = n$, we can use (4.48) and obtain

$$c_n = \frac{1}{n!}\sum_{\|\alpha\|=n}(-1)^{\alpha_2+\alpha_4+\cdots}h_\alpha s_\alpha \tag{4.66}$$

and

$$w_n = \frac{1}{n!}\sum_{\|\alpha\|=n}h_\alpha s_\alpha. \tag{4.67}$$

We resume the study of the Schur functions. Although ultimately $F^{(p)}$ has to be expressed in terms of $\mathscr{S}$, there are advantages in using $\mathscr{C}$ or $\mathscr{W}$ as an intermediate basis, because explicit formulae are then available. In fact as early as 1841 Jacobi had discovered the expression

$$\det(x_j^{p_i+n-i})/\det(x_j^{n-i}) = \det(w_{p_i-i+j}) \tag{4.68}$$

for the alternant quotient (Appendix, p. 188), which is precisely what Schur needed 67 years later. Thus denoting the determinant on the right by $W^{(p)}$ we have that

$$F^{(p)} = \det(w_{p_i-i+j}) = W^{(p)}. \tag{4.69}$$

126

The structure of $W^{(p)}$ is easily remembered: write the terms of p as suffixes attached to the diagonal elements; in any particular row the suffix increases by unity at each step as we move from the diagonal to the right and it decreases by unity as we move to the left. For example, when the partition is

$$4 = 2 + 1 + 1 + 0,$$

the determinant becomes

$$\begin{vmatrix} w_2 & w_3 & w_4 & w_5 \\ w_0 & w_1 & w_2 & w_3 \\ 0 & w_0 & w_1 & w_2 \\ 0 & 0 & 0 & w_0 \end{vmatrix}.$$

In (4.69) the determinant has n rows and columns, but the order is diminished when p involves zero terms. For if $p_n = 0$, the last row becomes $(0, 0, \ldots, 0, 1)$, as in the above example. Consequently the last row and the last column can be omitted. The procedure is repeated if $p_{n-1} = 0$, and so on. The result may be summarised as follows:

Proposition 4.5. *The Schur function which corresponds to the partition*

$$p: p_1 \geqslant p_2 \geqslant \ldots \geqslant p_u > 0, \quad p_{u+1} = \ldots = p_n = 0; \quad |p| = n$$

is given by

$$F^{(p)} = \det(w_{p_i - i + j}) \quad (i, j = 1, 2, \ldots, u). \tag{4.70}$$

A partition is often described by displaying its distinct positive parts, repetitions being indicated by an exponential notation. For example,

$$12 = 5 + 2 + 2 + 1 + 1 + 1$$

is written

$$[5, 2^2, 1^3].$$

Thus, for each n, we have that

$$[n]: n = n + 0 + \ldots + 0$$

and

$$[1^n]: n = 1 + 1 + \ldots + 1.$$

127

When $p = [n]$, the determinant (4.70) reduces to w_n. Hence, by (4.65),

$$F^{[n]} = w_n = \frac{1}{n!} \sum_\alpha h_\alpha s_\alpha.$$

Comparing this result with (4.60) we deduce that

$$\chi_\alpha^{[n]} = 1,$$

that is, $\chi^{[n]}$ is the trivial character.

When $p = [1^n]$, it is more convenient to use the original definition (4.58). Thus

$$F^{[1^n]} = \det(x_j^{n+1-i}) / \det(x_j^{n-i}).$$

On extracting the factor x_i from the ith row of the numerator, we find that

$$F^{[1^n]} = c_n,$$

whence by (4.60) and (4.66),

$$\chi_\alpha^{[1^n]} = (-1)^{\alpha_2 + \alpha_4 + \cdots},$$

which is the alternating character of S_n.

4.10. Conjugate partitions

When zero terms are disregarded, a partition

$$p : p_1 \geqslant p_2 \geqslant \ldots \geqslant p_u > 0, \quad |p| = n$$

may conveniently be described by a *graph* (p). This consists of p_1 nodes in the first row, p_2 nodes in the second row, $\ldots$, p_u nodes in the uth row, the initial nodes in each row being vertically aligned. For example, the partition

$$p : 4 + 2 + 2 + 1 + 1 + 1 = 11$$

is depicted by the graph

$$
\begin{array}{cccc}
\cdot & \cdot & \cdot & \cdot \\
\cdot & \cdot & & \\
\cdot & \cdot & & \\
\cdot & & & \\
\cdot & & & \\
\cdot & & &
\end{array}
$$

(p)

With every partition p we associate its *conjugate partition* p' obtained by interchanging rows and columns of the graph (p); in other words the terms of p' are the number of nodes in the columns of (p). Evidently, the conjugate of p' is p. (This notion of conjugacy is not related to the same term used in group theory.) In the above example

$$p': 6 + 3 + 1 + 1 = 11.$$

A partition which is identical with its conjugate is said to be *self-conjugate*. For example,

$$4 + 2 + 1 + 1 = 8$$

is self-conjugate, as is seen from its graph, namely

By scanning the columns of (p) from right to left we find that the terms of p' are represented by the symbol

$$p' = [1^{p_1 - p_2}, 2^{p_2 - p_3}, 3^{p_3 - p_4}, \ldots, u^{p_u}],$$

with the understanding that no term of p' is equal to j if $p_j = p_{j+1}$. Whereas the graph furnishes an intuitive definition of conjugate partitions, it is desirable to have an alternative characterisation which is based solely on numerical relationships. This is expressed in the following lemma.

Lemma. *The partitions*

$$p: p_1 \geqslant p_2 \geqslant \ldots \geqslant p_u > 0, \quad |p| = n$$

and

$$q: q_1 \geqslant q_1 \geqslant \ldots \geqslant q_v > 0, \quad |q| = n$$

are conjugate if and only if

$$p_1 = v, \qquad q_1 = u \tag{4.71}$$

and

$$p_i + q_j \neq i + j - 1 \tag{4.72}$$

$(i = 1, 2, \ldots, u; j = 1, 2, \ldots, v)$.

129

Proof. (i) Suppose that p and q are conjugate. Select a pair of integers (i, j) satisfying $1 \leqslant i \leqslant u$, $1 \leqslant j \leqslant v$. Two cases arise:

(a) If $p_i \geqslant j$, then the jth column of the graph (p) extends at least down to the ith row, that is $q_j \geqslant i$. Hence

$$p_i + q_j \geqslant j + i > i + j - 1.$$

(b) If $p_i < j$, then the jth column of (p) terminates before the ith row, that is $q_j < i$. Thus we have that $p_i \leqslant j - 1$ and $q_j \leqslant i - 1$, whence

$$p_i + q_j \leqslant i + j - 2 < i + j - 1.$$

Therefore the inequality (4.72) holds in all cases. The equations (4.71) are an immediate consequence of the geometric definition of conjugacy.

(ii) Conversely, suppose the conditions (4.71) and (4.72) are satisfied by the non-decreasing partitions p and q. We claim that the numbers

$$p_i + u - i \quad (i = 1, 2, \ldots, u) \tag{4.73}$$

and

$$u - 1 + j - q_j \quad (j = 1, 2, \ldots, v) \tag{4.74}$$

are distinct and together constitute the set

$$0, 1, 2, \ldots, u + v - 1 \tag{4.75}$$

in some order. First, since the terms of p are non-increasing, we have that

$$p_1 + u - 1 > p_2 + u - 2 > p_3 + u - 3 > \ldots > p_u.$$

Also, since $p_u \geqslant 1$,

$$1 \leqslant p_u + u - u \leqslant p_i + u - i \leqslant p_1 + u - 1,$$

that is,

$$0 < p_i + u - i \leqslant v + u - 1.$$

Hence (4.73) consists of u distinct members of (4.75). Similarly, the integers (4.74) are distinct and, since $q_v \geqslant 1$,

$$0 = u - 1 + 1 - q_1 \leqslant u - 1 + j - q_j \leqslant u - 1 + v - q_v \leqslant u + v - 2,$$

that is

$$0 \leqslant u - 1 + j - q_j \leqslant u + v - 2.$$

Thus (4.74) is a subset of (4.75). Next, we observe that (4.73) and (4.74) are disjoint by virtue of (4.72); in other words, they are complementary subsets of (4.72), as asserted.

It follows that, when p is given, the set (4.74) and consequently the integers $q_1, q_2, \ldots, q_v$ are uniquely determined by the conditions (4.71) and (4.72). Without loss of generality we may assume that

$$q : q_1 \geqslant q_2 \geqslant \ldots \geqslant q_v.$$

Moreover, q is a partition of n, because we have that

$$\sum_{i=1}^{u} (p_i + u - i) + \sum_{j=1}^{v} (u - 1 + j - q_j) = \sum_{k=0}^{u+v-1} k,$$

which after a short calculation is seen to be equivalent to

$$\sum_i p_i = \sum_j q_j.$$

We conclude that because of its uniqueness q is identical with the conjugate of p', which is known to satisfy (4.71) and (4.72). This proves the lemma.

The notion of conjugacy can be extended to partitions with zero terms by stipulating that two partitions are conjugate if and only if they are conjugate in the previous sense after the zero terms have been removed. The preceding lemma can be adapted to provide an arithmetical criterion for conjugacy, which will play a crucial rôle later on (p. 134).

Proposition 4.6. *The partitions*

$$p : p_1 \geqslant p_2 \geqslant \ldots \geqslant p_n \geqslant 0, \quad |p| = n$$

and

$$q : q_1 \geqslant q_2 \geqslant \ldots \geqslant q_n \geqslant 0, \quad |q| = n$$

are conjugate if and only if

$$p_i + n - i \quad (i = 1, 2, \ldots, n) \tag{4.76}$$

and

$$n - 1 + j - q_j \quad (j = 1, 2, \ldots, n) \tag{4.77}$$

are complementary subsets of

$$0, 1, 2, \ldots, 2n - 1. \tag{4.78}$$

Proof. (i) Suppose that p and q are conjugate partitions. Displaying the zero terms explicitly we write

$$p_1 \geqslant p_2 \geqslant \ldots \geqslant p_u > 0 \qquad p_{u+1} = \ldots = p_n = 0 \tag{4.79}$$

and

$$q_1 \geq q_2 \geq \ldots \geq q_v > 0 \qquad q_{v+1} = \ldots = q_n = 0, \tag{4.80}$$

with the understanding that the equations (4.79) or (4.80) are modified if $u = n$ or $v = n$ respectively. By the lemma we have that $p_1 = v$, $q_1 = u$ and

$$p_i + q_j \neq i + j - 1 \tag{4.81}$$

provided that $1 \leq i \leq u$ and $1 \leq j \leq v$. We shall show that the inequalities (4.81) continue to hold when $p_i = 0$ or $q_j = 0$. For if $p_i = 0$, then $i > u$ and

$$p_i + q_j = q_j \leq q_1 = u < i \leq i + j - 1.$$

Next, if $q_j = 0$, then $j > v$ and

$$p_i + q_j = p_i \leq p_1 = v < j \leq i + j - 1.$$

Thus (4.81) is true in all cases. Also

$$0 \leq p_i + n - i \leq 2n - 1$$

and

$$0 \leq n - 1 + j - q_j \leq 2n - 1.$$

The inequalities ensure that (4.76) and (4.77) are complementary subsets of (4.78).

(ii) Conversely, if the conditions of this proposition are satisfied, the sets (4.76) and (4.77) determine each other uniquely. In particular when p is given we can find q_j $(j = 1, 2, \ldots, n)$ from (4.77), and we may assume that

$$q_1 \geq q_2 \geq q_3 \geq \ldots \geq q_n.$$

On summing the members of (4.76) and (4.77) we obtain that

$$\sum_{i=1}^{n} (p_i + n - i) + \sum_{j=1}^{n} (n - 1 + j - q_j) = \sum_{k=0}^{2n-1} k,$$

which readily reduces to

$$\sum_i p_i = \sum_j q_j.$$

We conclude that there is a unique partition q which together with p fulfils the conditions of this proposition. Hence q must be identical with the conjugate of p, which is known to satisfy those conditions.

We proceed to examine the relationship between the Schur functions that correspond to two conjugate partitions. To this end we have to bring into play the set $\mathscr{C}$; but it is expedient temporarily to put

$$a_m = (-1)^m c_m \quad (m = 0, \pm 1, \pm 2, \dots). \tag{4.82}$$

The equations (4.63) can then be written as

$$a_0 w_m + a_1 w_{m-1} + \dots + a_n w_{m-n} = \delta_{0,m}. \tag{4.83}$$

We now introduce the $2n \times 2n$ matrix

$$W = (w_{s-r}) = \begin{vmatrix} w_0 & w_1 & w_2 & \dots & w_{2n-2} & w_{2n-1} \\ 0 & w_0 & w_1 & \dots & w_{2n-3} & w_{2n-2} \\ \dots & \dots & \dots & & \dots & \dots \\ 0 & 0 & 0 & \dots & w_0 & w_1 \\ 0 & 0 & 0 & \dots & 0 & w_0 \end{vmatrix}, \tag{4.84}$$

where r and s are the row and column suffixes respectively, each running from 0 to $2n-1$. The significance of W stems from the fact that every Schur function of S_n can be expressed as an n-rowed subdeterminant (minor) of W. Generally, we denote by

$$W(\lambda; \mu) \tag{4.85}$$

the submatrix of W whose elements stand at the intersections of the rows and columns labelled by

$$\lambda : 0 \leqslant \lambda_1 < \lambda_2 < \dots < \lambda_n \leqslant 2n - 1$$

and

$$\mu : 0 \leqslant \mu_1 < \mu_2 < \dots < \mu_n \leqslant 2n - 1$$

respectively. Let

$$p : p_1 \geqslant p_2 \geqslant \dots \geqslant p_n \geqslant 0, \quad |p| = n$$

be an arbitrary partition of n. Then

$$p_n < p_{n-1} + 1 < p_{n-2} + 2 < \dots < p_1 + n - 1.$$

Hence we may put

$$\lambda_i = i - 1; \qquad \mu_j = p_{n+1-j} + (j-1), \tag{4.86}$$

$(i, j = 1, 2, \dots, n)$. The (i, j)th element of (4.85) is then obtained by putting $r = \lambda_i$, $s = \mu_j$. Thus

$$W(\lambda; \mu) = (w_{p_{n+1-j}+j-i}).$$

133

Since we are only interested in the determinant of this matrix, we may simultaneously reverse the order of the rows and of the columns. This does not alter the value of the determinant, as the rows and columns undergo the same permutation. Thus we replace i by $n+1-i$ and j by $n+1-j$ and obtain that

$$\det W(\lambda\,;\mu) = \det(w_{p_j-j+i}).$$

Finally, we transpose the determinant, that is we interchange i and j, which again does not affect its value. Thus, by (4.69),

$$\det W(\lambda\,;\mu) = \det(w_{p_i-i+j}) = F^{(p)}. \tag{4.87}$$

In analogy with (4.84) we construct the matrix

$$A = (a_{s-r}) \quad (r,s = 0, 1, \ldots, 2n-1).$$

Then the (r,s)th element of AW is given by

$$\sum_{t=0}^{2n-1} a_{t-r}w_{s-t} = a_0 w_{s-r} + a_1 w_{s-r-1} + \ldots + a_n w_{s-r-n}.$$

By (4.83) this expression is equal to $\delta_{0,s-r}$, that is $\delta_{s,r}$. Hence A and W are inverse to each other, and we may apply Jacobi's Theorem (Appendix, p. 191) in relation to their n-rowed minors. Referring to (4.86) we observe that the set of indices complementary to λ in

$$0, 1, 2, \ldots, 2n-1$$

is given by

$$\rho: \rho_i = n+i-1 \quad (i = 1, 2, \ldots, n),$$

while the complementary set of μ consists of

$$\sigma: \sigma_i = n-1+i-q_i \quad (i = 1, 2, \ldots, n)$$

by virtue of Proposition 4.6, where q is the conjugate partition of p. The matrices A and W are in upper triangular form, and

$$\det A = \det W = 1.$$

According to Jacobi's Theorem we have that

$$\det W(\lambda\,;\mu) = (-1)^\varepsilon \det A(\sigma\,;\rho), \tag{4.88}$$

where

$$\varepsilon = \sum_i \lambda_i + \sum_i \mu_i \equiv \sum_j p_j = n \pmod 2.$$

The (i, j)th element of $A(\sigma; \rho)$ is equal to a_{s-r} where $r = \sigma_i$ and $s = \rho_j$. Thus

$$\det A(\sigma; \rho) = \det(a_{q_i-i+j}).$$

Next, we reintroduce the functions c_i by means of (4.82). Extracting the minus signs from all the elements of the determinant we obtain that

$$\det A(\sigma; \rho) = (-1)^{\varepsilon'} \det(c_{q_i-i+j}),$$

where

$$\varepsilon' = \sum_i (q_i - i) + \sum_j j = n.$$

Hence the sign in (4.88) is cancelled, and we have established the remarkable formula

$$\det(w_{p_i-i+j}) = \det(c_{q_i-i+j}). \tag{4.89}$$

It remains to express this result in terms of the system $\mathscr{S}$. We recall (p. 126) that each function w is changed into the corresponding c when $s_2, s_4 \ldots$ are replaced by $-s_2, -s_4, \ldots$. Hence from the expansion

$$F^{(q)} = \det(w_{q_i-i+j}) = \frac{1}{n!} \sum_\alpha h_\alpha \chi_\alpha^{(q)} \mathbf{s}_\alpha$$

we deduce that

$$F^{(p)} = \det(c_{q_i-i+j}) = \frac{1}{n!} \sum_\alpha (-1)^{\alpha_2+\alpha_4+\cdots} h_\alpha \chi_\alpha^{(q)} \mathbf{s}_\alpha. \tag{4.90}$$

We summarise this result as follows:

Theorem 4.3. *If p and q are conjugate partitions of n, then*

$$\chi^{(p)} = \chi^{[1^n]} \chi^{(q)}. \tag{4.91}$$

4.11. The characters of S_5

We apply the foregoing theory to S_5. This is presented merely as an illustration; the group S_5 is small enough to be amenable to a more direct treatment.

In contrast to the labelling of characters it is customary to write the terms of a partition in ascending order when specifying the cycle pattern of a conjugacy class. For example, the symbol

$$(1^3 \, 2^2 \, 5)$$

denotes the class of S_{12} whose members are the product of three 1-cycles, two 2-cycles and one 5-cycle. The classes of S_5 and their sizes are exhibited in Table 4.2,

α	(1^5)	$(1^3 2)$	$(1^2 3)$	$(1\,2^2)$	(14)	(23)	(5)
h_α	1	10	20	15	30	20	24

Table 4.2.

h_α may be computed from (4.10). Henceforth we shall enumerate the classes in the order given in Table 4.2. By (4.67) the same table may be used to express w_5 in terms of $\mathscr{S}$, thus

$$w_5 = \tfrac{1}{120}(s_1^5 + 10s_1^3 s_2 + 20s_1^2 s_3 + 15s_1 s_2^2 + 30s_1 s_4 + 20s_2 s_3 + 24s_5).$$

Similarly, by referring to the class sizes of S_m ($m = 1, 2, 3, 4$) (see Tables 2.2 and 4.1, pp. 50 and 106), we find that

$$w_1 = s_1,$$
$$w_2 = \tfrac{1}{2}(s_1^2 + s_2),$$
$$w_3 = \tfrac{1}{6}(s_1^3 + 3s_1 s_2 + 2s_3),$$
$$w_4 = \tfrac{1}{24}(s_1^4 + 6s_1^2 s_2 + 8s_1 s_3 + 3s_2^2 + 6s_4).$$

Alternatively, these expressions for w_r ($1 \leqslant r \leqslant 5$) can be obtained by means of the recurrence relation or by the determinants mentioned in Exercises 4, 5 and 6. We infer that S_5 possesses seven simple characters. Two of these, namely $\chi^{[5]}$ and $\chi^{[1^5]}$, have already been identified. The remaining five characters, or their conjugates, will be generated by Schur functions. By (4.70),

$$F^{[41]} = \begin{vmatrix} w_4 & w_5 \\ w_0 & w_1 \end{vmatrix} = w_4 w_1 - w_5$$

$$= \tfrac{1}{120}\{4s_1^5 + 20s_1^3 s_2 + 20s_1^2 s_3 - 20s_2 s_3 - 24s_5\}.$$

Hence by (4.60) and Table 4.2, we find that

$$\chi^{[41]}: (4, 2, 1, 0, 0, -1, -1).$$

136

Next,

$$F^{[32]} = \begin{vmatrix} w_3 & w_4 \\ w_1 & w_2 \end{vmatrix} = w_3 w_2 - w_4 w_1$$

$$= \tfrac{1}{120}\{5s_1^5 + 10s_1^3 s_2 - 20s_1^2 s_3 + 15s_1 s_2^2 - 30s_1 s_4 + 20s_2 s_3\}$$

whence

$$\chi^{[32]}: (5, 1, -1, 1, -1, 1, 0).$$

Similarly,

$$F^{[3\,1^2]} = \begin{vmatrix} w_3 & w_4 & w_5 \\ w_0 & w_1 & w_2 \\ 0 & w_0 & w_1 \end{vmatrix} = w_3 w_1^2 + w_5 - w_3 w_2 - w_4 w_1$$

$$= \tfrac{1}{120}\{6s_1^5 - 30s_1 s_2^2 + 24s_5\},$$

which yields

$$\chi^{[3\,1^2]}: (6, 0, 0, -2, 0, 0, 1).$$

An inspection of the graphs shows that $[2\,1^3]$ is conjugate to $[41]$ and that $[2^2\,1]$ is conjugate to $[32]$, while $[3\,1^2]$ is self-conjugate. Hence, by (4.91),

$$\chi^{[2\,1^3]} = \chi^{[1^5]} \chi^{[41]}$$

and

$$\chi^{[2^2\,1]} = \chi^{[1^5]} \chi^{[32]}.$$

Thus we have found all seven characters. They are listed in Table 4.3.

$$S_5:$$

α	(1^5)	$(1^3 2)$	$(1^2 3)$	$(1\,2^2)$	(14)	(23)	(5)
h_α	1	10	20	15	30	20	24
$\chi^{[5]}$	1	1	1	1	1	1	1
$\chi^{[41]}$	4	2	1	0	0	-1	-1
$\chi^{[32]}$	5	1	-1	1	-1	1	0
$\chi^{[3\,1^2]}$	6	0	0	-2	0	0	1
$\chi^{[2^2\,1]}$	5	-1	-1	1	1	-1	0
$\chi^{[2\,1^3]}$	4	-2	1	0	0	1	-1
$\chi^{[1^5]}$	1	-1	1	1	-1	-1	1

Table 4.3.

Exercises

1. After a jolly party each reveller picks up at random one of the umbrellas which had previously been deposited in the cloakroom. Prove that only one gentleman is expected to return with his own umbrella.

[Randomness here means that all permutations of the umbrellas are equally likely, and the expected value is the average.]

2. Let $F^{(p)}$ be the Schur function corresponding to the partition

$$p: p_1 \geqslant p_2 \geqslant \ldots \geqslant p_n \geqslant 0.$$

Referring to (4.58) regard $F^{(p)}$ as a polynomial in $x_1, x_2, \ldots, x_n$. Denote by $F_m^{(p)}$ the polynomial obtained by putting

$$x_{m+1} = x_{m+2} = \ldots = x_n = 0 \quad (m < n).$$

Show that

(i) $F_m^{(p)} = 0$, if $p_{m+1} > 0$,

and

(ii) $F_m^{(p)} = \det(x_j^{p_i+m-i}) / \det(x_j^{m-i})$, if $p_{m+1} = 0$

$(i, j = 1, 2, \ldots, m)$.

3. Show that the degree of the irreducible representation of S_n that corresponds to the partition

$$p: p_1 \geqslant p_2 \geqslant \ldots \geqslant p_m \geqslant 0, \quad p_{m+1} = \ldots = p_n = 0 \quad (m < n)$$

is equal to

$$f^{(p)} = \frac{n! \prod_{i<j}(p_i - p_j + j - i)}{\prod_i (p_i + m - i)!}$$

$(i, j = 1, 2, \ldots, m)$.

4. Show that the characters of S_n that correspond to partitions with at most two positive terms are given by the generating function

$$(1-x) \prod_{r=1}^{n} (1+x^r)^{\alpha_r} = \sum_{t=0}^{[\frac{1}{2}n]} \chi_\alpha^{[n-t,t]}(x^t - x^{n+1-t}),$$

where

$$\alpha: \alpha_1 + 2\alpha_2 + \ldots + n\alpha_n = n.$$

(In the above formula, $[n, 0]$ is to be identified with $[n]$.) In particular, verify that

(i) $\chi_\alpha^{[n-1,1]} = \alpha_1 - 1 \quad (n \geqslant 2)$,

(ii) $\chi_\alpha^{[n-2,2]} = \frac{1}{2}\alpha_1(\alpha_1 - 3) + \alpha_2 \quad (n \geqslant 4)$,

(iii) $\chi_\alpha^{[n-3,3]} = \frac{1}{6}\alpha_1(\alpha_1 - 1)(\alpha_1 - 5) + (\alpha_1 - 1)\alpha_2 + \alpha_3 \quad (n \geqslant 6)$.

5. Let
$$f(t) = \prod_{i=1}^{n} (1 - x_i t) = 1 - c_1 t + c_2 t^2 - \ldots + (-1)^n c_n.$$

Show that
$$\frac{f'(t)}{f(t)} = -(s_1 + s_2 t + s_3 t^2 + \ldots).$$

Deduce the recurrence relation (*Newton's formulae*)
$$s_r - s_{r-1} c_1 + s_{r-2} c_2 - \ldots + (-1)^{r-1} s_1 c_{r-1} + (-1)^r r c_r = 0$$

and prove that

$$c_r = \frac{1}{r!} \begin{vmatrix} s_1 & 1 & 0 & \ldots & 0 & 0 \\ s_2 & s_1 & 2 & \ldots & 0 & 0 \\ \ldots & \ldots & \ldots & & \ldots & \ldots \\ s_{r-1} & s_{r-2} & s_{r-3} & \ldots & s_1 & r-1 \\ s_r & s_{r-1} & s_{r-2} & \ldots & s_2 & s_1 \end{vmatrix}$$

$(r = 1, 2, \ldots, n)$.

6. Establish the recurrence relation (*Brioschi's formulae*)
$$s_r + s_{r-1} w_1 + s_{r-2} w_2 + \ldots + s_1 w_{r-1} - r w_r = 0,$$

and prove that

$$w_r = \frac{1}{r!} \begin{vmatrix} s_1 & -1 & 0 & \ldots & 0 & 0 \\ s_2 & s_1 & -2 & \ldots & 0 & 0 \\ \ldots & \ldots & \ldots & & \ldots & \ldots \\ s_{r-1} & s_{r-2} & s_{r-3} & \ldots & s_1 & -r+1 \\ s_r & s_{r-1} & s_{r-2} & \ldots & s_2 & s_1 \end{vmatrix}$$

$(r = 1, 2, \ldots, n)$.

7. Show that if
$$g(\alpha) = g(\alpha_1, \alpha_2, \ldots) = (1^{\alpha_1} \alpha_1! 2^{\alpha_2} \alpha_2! \ldots)^{-1},$$

then
$$\frac{g(\alpha_1 - 2, \alpha_2, \alpha_3, \ldots)}{g(\alpha_1, \alpha_2, \alpha_3, \ldots)} = \alpha_1(\alpha_1 - 1),$$

$$\frac{g(\alpha_1, \alpha_2 - 1, \alpha_3, \ldots)}{g(\alpha_1, \alpha_2, \alpha_3, \ldots)} = 2\alpha_2.$$

Prove that, if $n \geq 3$,
$$\chi_\alpha^{[n-2, 1^2]} = \tfrac{1}{2}(\alpha_1 - 1)(\alpha_1 - 2) - \alpha_2.$$

8. The number of terms of the partition

$$p: p_1 \geqslant p_2 \geqslant \ldots \geqslant p_n \geqslant 0$$

which satisfy $p_i \geqslant i$ is called the *rank* of p. Show that the partitions of rank unity are given by

$$[n-t, 1^t] \quad (t = 0, 1, \ldots, n-1)$$

provided that this symbol is interpreted as $[n]$ when $t = 0$ and as $[1^n]$ when $t = n-1$.

9. Suppose that the members of C_α are products of t cycles. Prove that

$$\chi_\alpha^{(p)} = 0$$

if the rank of p is greater than t.

10. Let C_γ be the conjugacy class of S_n which contains the full cycle $(1\,2\ldots n)$. Prove that

$$\chi_\gamma^{(p)} = 0,$$

unless

$$p = [n-t, 1^t] \quad (t = 0, 1, \ldots, n-1),$$

in which case

$$\chi_\gamma^{(p)} = (-1)^t.$$

11. Construct the character table for the group S_6 given in Table 4.4.

$$S_6:$$

C_α	1^6	$1^4 2$	$1^3 3$	$1^2 2^2$	$1^2 4$	123	15	2^3	24	3^2	6
h_α	1	15	40	45	90	120	144	15	90	40	120
[6]	1	1	1	1	1	1	1	1	1	1	1
[51]	5	3	2	1	1	0	0	-1	-1	-1	-1
[42]	9	3	0	1	-1	0	-1	3	1	0	0
[4 1²]	10	2	1	-2	0	-1	0	-2	0	1	1
[3²]	5	1	-1	1	-1	1	0	-3	-1	2	0
[321]	16	0	-2	0	0	0	1	0	0	-2	0
[3 1³]	10	-2	1	-2	0	1	0	2	0	1	-1
[2³]	5	-1	-1	1	1	-1	0	3	-1	2	0
[2² 1²]	9	-3	0	1	1	0	-1	-3	1	0	0
[2 1⁴]	5	-3	2	1	-1	0	0	1	-1	-1	1
[1⁶]	1	-1	1	1	-1	-1	1	-1	1	1	-1

Table 4.4.

Note: For ease of printing the character $\chi^{(p)}$ is here simply written as $[p]$.

5

GROUP-THEORETICAL APPLICATIONS

5.1. Algebraic numbers

In this section we assemble a few basic facts about algebraic numbers, which will be required in the sequel. For a more detailed exposition the reader should consult one of the standard texts [e.g. **18**]. All numbers considered in this chapter lie in the complex field. As usual, the field of rational numbers will be denoted by $\mathbb{Q}$.

Definition 5.1. *A number θ is called algebraic if it satisfies an equation of the form*

$$\theta^m + b_1 \theta^{m-1} + \ldots + b_{m-1}\theta + b_m = 0,$$

where $b_1, \ldots, b_{m-1}, b_m$ lie in $\mathbb{Q}$.

Without loss of generality, we have taken the coefficient of the highest term to be equal to unity. Thus θ is algebraic if it is the root of a monic polynomial in $\mathbb{Q}[x]$.

Clearly, if θ is algebraic, it must satisfy a monic equation of least degree n, say $M(\theta) = 0$, where

$$M(x) = x^n + a_1 x^{n-1} + \ldots + a_n \tag{5.1}$$

lies in $\mathbb{Q}[x]$. This equation is uniquely determined; for if θ were also a root of

$$x^n + a_1' x^{n-1} + \ldots + a_n',$$

then on subtracting we should find that θ is a root of

$$(a_1 - a_1')x^{n-1} + \ldots + (a_n - a_n'),$$

which contradicts the minimality of n, unless $a_i = a_i'$ $(i = 1, \ldots, n)$. Thus with every algebraic number there is associated a unique *monic minimal polynomial* in $\mathbb{Q}[x]$. Evidently, this minimal polynomial is irreducible in $\mathbb{Q}[x]$.

Let the roots of $M(x)$ be the (complex) numbers

$$\theta, \theta', \theta'', \ldots, \theta^{(n-1)}. \tag{5.2}$$

141

Thus

$$M(\theta) = M(\theta') = \ldots = M(\theta^{(n-1)}) = 0.$$

The numbers (5.2) are said to form a set of *algebraic conjugates*. In other words, algebraic conjugates share the same minimal polynomial

$$M(x) = \prod_{i=0}^{n-1} (x - \theta^{(i)}) \quad (\theta^{(0)} = \theta).$$

Suppose now that $L(x)$ is any polynomial in $\mathbb{Q}[x]$ such that $L(\theta) = 0$. On dividing L by M we have that

$$L(x) = M(x)Q(x) + R(x), \tag{5.3}$$

where the degree of R is less than n. Substituting $x = \theta$ in (5.3) we find that $R(\theta) = 0$. This contradicts the minimality of n, unless R is the zero polynomial. Hence

$$L(x) = M(x)Q(x),$$

and it follows that $L(\theta^{(i)}) = 0$ $(i = 0, 1, \ldots, n-1)$. Thus we have

Proposition 5.1. *Let θ be an algebraic number. If $L(x)$ is any polynomial in $\mathbb{Q}[x]$ which has θ as a root, then each conjugate of θ is also a root of $L(x)$.*

Next, we consider numbers of the form

$$\alpha = T(\theta) = t_0 + t_1\theta + \ldots + t_{n-1}\theta^{n-1},$$

where $t_0, t_1, \ldots, t_{n-1}$ are rational. Then

$$P(x) = \prod_{i=0}^{n-1} \left(x - T(\theta^{(i)})\right) \tag{5.4}$$

is a polynomial in x of degree n. Its coefficients are symmetric functions of the set (5.2) with rational coefficients; for example, the coefficient of x^{n-1} in P is

$$-\left(T(\theta) + T(\theta') + \ldots + T(\theta^{(n-1)})\right).$$

By the Fundamental Theorem on symmetric functions this and all the other coefficients of P are rationally expressible in terms of the *elementary symmetric functions* [18, p. 32]

$$-a_1 = \theta + \theta' + \ldots + \theta^{(n-1)},$$

$$a_2 = \theta\theta' + \theta\theta'' + \ldots + \theta^{(n-2)}\theta^{(n-1)},$$

$$\vdots$$

$$(-1)^n a_n = \theta\theta' \ldots \theta^{(n-1)},$$

where $a_1, a_2, \ldots, a_n$ are the coefficients of $M(x)$, which are rational. Hence $P(x)$ lies in $\mathbb{Q}[x]$. By construction, we have that $P(\alpha) = 0$. This shows immediately that α is algebraic. But P need not be the minimal polynomial for α. However, if

$$\alpha, \alpha', \alpha'', \ldots \tag{5.5}$$

are the conjugates of α, then Proposition 5.1 tells us that each of the numbers (5.5) is a root of P, which are displayed in (5.4). Hence we have

Proposition 5.2. *Suppose that θ is algebraic, and let $\alpha = T(\theta)$, where $T(x) \in \mathbb{Q}[x]$. Then α is algebraic and its conjugates are found amongst the numbers*

$$T(\theta), T(\theta'), \ldots . T(\theta^{(n-1)}).$$

Definition 5.2. *A number ξ is called an algebraic integer if it satisfies an equation of the form*

$$\xi^m + b_1 \xi^{m-1} + \ldots + b_m = 0, \tag{5.6}$$

where $b_1, \ldots, b_m$ are integers.

In this definition it is not assumed that (5.6) is the minimal equation for ξ. For it can be shown that, if ξ satisfies any equation of the form (5.6), then its monic minimal polynomial is bound to have integral coefficients. This fact is based on a result of Gauss, which we here quote without proof [**18**, p. 27]:

Theorem 5.1. *The minimal polynomial of an algebraic integer is necessarily of the form*

$$x^n + a_1 x^{n-1} + \ldots + a_n = 0,$$

where $a_1, \ldots, a_n$ are integers.

The familiar integers $0, \pm 1, \pm 2, \ldots, \pm k, \ldots$ are now often referred to as *rational integers*; they are a particular kind of algebraic integer, as they satisfy an equation of the form

$$\xi + b = 0$$

in accordance with Definition 5.2.

In order to prove that a certain number is an algebraic integer we may avail ourselves of the following simple device.

143

Lemma. *Let $y_1, y_2, \ldots, y_N$ be complex numbers, not all zero, and suppose that ξ satisfies equations of the form*

$$\xi y_i = \sum_{j=1}^{N} a_{ij} y_j \quad (i = 1, 2, \ldots, N), \tag{5.7}$$

where the a_{ij} are rational integers. Then ξ is an algebraic integer.

Proof. We may regard (5.7) as a system of linear homogeneous equations for $y_1, y_2, \ldots, y_N$. Since, by hypothesis, this system has a non-zero solution, the determinant of the coefficients must vanish, that is

$$\det(\xi \delta_{ij} - a_{ij}) = 0.$$

This is a monic equation for ξ (of degree N) with integral coefficients. Hence ξ is an algebraic integer.

As an application of this method we prove the following

Theorem 5.2. *If ξ and η are algebraic integers, so are $\xi + \eta$ and $\xi\eta$.*

Proof. Slightly modifying Definition 5.2 we can say that ξ is an algebraic integer if it satisfies an equation of the form

$$\xi^r = a_1 \xi^{r-1} + \ldots + a_{r-1}\xi + a_r \quad (a_i \text{ integers}). \tag{5.8}$$

We claim that every power ξ^p ($p \geqslant 0$) can be written as a linear combination of

$$1, \xi, \xi^2, \ldots, \xi^{r-1}$$

with integral coefficients. This is trivial when $0 \leqslant p \leqslant r - 1$; when $p = r$, the assertion is proved by (5.8). Next, suppose that $p = r + 1$. Multiply (5.8) throughout by ξ, thus

$$\xi^{r+1} = a_1 \xi^r + a_2 \xi^{r-1} + \ldots + a_{r-1}\xi^2 + a_r \xi.$$

On eliminating ξ^r on the right with the aid of (5.8) we obtain an equation of the form

$$\xi^{r+1} = a_1' \xi^{r-1} + a_2' \xi^{r-2} + \ldots + a_r',$$

where $a_1', a_2', \ldots$ are integers. Proceeding in this way we establish the result for an arbitrary power ξ^p. Similarly, it may be assumed that η satisfies an equation of the form

$$\eta^s = b_1 \eta^{s-1} + \ldots + b_{s-1}\eta + b_s \quad (b_j \text{ integers}),$$

whence every power $\eta^q (q \geqslant 0)$ can be expressed as a linear combination of

$$1, \eta, \ldots, \eta^{s-1}$$

with integral coefficients. Let $y_1, y_2, \ldots, y_N$ be the products

$$\xi^i \eta^j \quad (i = 0, 1, \ldots, r-1; j = 0, 1, \ldots, s-1) \tag{5.9}$$

arranged in some fixed order. From the foregoing observations it follows that any product of the form $\xi^p \eta^q$ can be expressed linearly in terms of (5.9), that is, in terms of $y_1, y_2, \ldots, y_N$, with integral coefficients. Hence there are equations

$$\left.\begin{aligned}
(\xi + \eta) y_i &= \sum_{j=1}^{N} c_{ij} y_j \\
\xi \eta y_i &= \sum_{j=1}^{N} d_{ij} y_j
\end{aligned}\right\} \quad (i = 1, 2, \ldots, N),$$

where c_{ij} and d_{ij} are integers. This proves that $\xi + \eta$ and $\xi \eta$ are algebraic integers.

Next we have to establish an elementary but important fact.

Theorem 5.3. *If ξ is both a rational number and an algebraic integer, then ξ is a rational integer.*

Proof. Let

$$\xi = \frac{r}{s} \tag{5.10}$$

be a fraction in its lowest terms, that is

$$(r, s) = 1 \tag{5.11}$$

and $s > 0$. By hypothesis, ξ satisfies an equation

$$\xi^m + b_1 \xi^{m-1} + \ldots + b_m = 0$$

with integral coefficients. Substituting (5.10) and clearing the denominators we obtain that

$$r^m = -s(b_1 r^{m-1} + b_2 s r^{m-2} + \ldots b_m s^{m-1}).$$

This is incompatible with (5.11), unless $s = 1$. Hence $\xi = r$, a rational integer.

145

For any positive integer g, the gth roots of unity are algebraic integers, because they are the roots of the polynomial

$$x^g - 1 = 0, \qquad (5.12)$$

which, of course, is not the minimal polynomial when $g > 1$. If we put

$$\varepsilon = \exp(2\pi i/g),$$

then the roots of (5.12) are ε^j $(j = 0, 1, \ldots, g - 1)$. At this stage, it is not essential to know the minimal polynomial for ε. But Proposition 5.1 tells us that the conjugates of ε also satisfy (5.12). Thus each conjugate of ε is of the form ε^j for some j. As a matter of fact, they are those ε^j for which $(j, g) = 1$.

Since the values of the characters are sums of roots of unity (p. 50), an application of Theorem 5.2 yields the following result.

Proposition 5.3. *All values of group characters are algebraic integers.*

5.2. Representations of the group algebra

We recall the notion of the group algebra, which we introduced in § 2.2 (p. 42): starting from a multiplicative group

$$G: x_1, x_2, \ldots, x_g$$

we constructed the algebra $G_{\mathbb{C}}$. Its elements are all formal sums

$$u = \alpha_1 x_1 + \alpha_2 x_2 + \ldots + \alpha_g x_g, \qquad (5.13)$$

for which addition and multiplication were defined in the obvious manner. We also studied the centre Z of $G_{\mathbb{C}}$: the elements of Z are uniquely expressed as

$$z = \zeta_1 c_1 + \zeta_2 c_2 + \ldots + \zeta_k c_k, \qquad (5.14)$$

where

$$c_\alpha = \sum_{x \in C_\alpha} x \quad (\alpha = 1, 2, \ldots, k) \qquad (5.15)$$

and $\zeta_1, \zeta_2, \ldots, \zeta_k$ are arbitrary complex numbers.

Evidently the product of two central elements lies again in the centre; in other words, Z is a subalgebra of $G_{\mathbb{C}}$. In particular $c_\alpha c_\beta$ lies in Z. Hence there are equations of the form

$$c_\alpha c_\beta = \sum_{\gamma=1}^{k} a_{\alpha\beta\gamma} c_\gamma \quad (\alpha, \beta = 1, 2, \ldots, k). \qquad (5.16)$$

It is easy to see that the coefficients $a_{\alpha\beta\gamma}$ are non-negative integers. For, by analogy with (5.15), let

$$c_\beta = \sum_{y \in C_\beta} y, \qquad c_\gamma = \sum_{z \in C_\gamma} z.$$

Then, if z_0 is a fixed element of C_γ, the coefficient $a_{\alpha\beta\gamma}$ is equal to the number of solutions of $z_0 = xy$, where $x \in C_\alpha$ and $y \in C_\beta$.

Suppose that A is a representation of G of degree m. We can extend A by linearity to G_C by defining

$$A(u) = \alpha_1 A(x_1) + \alpha_2 A(x_2) + \ldots + \alpha_g A(x_g),$$

where u is given in (5.13). It is clear that this will furnish a matrix representation of the algebra G_C, that is

$$A(u + v) = A(u) + A(v),$$

$$A(u)A(v) = A(uv),$$

$$A(\lambda u) = \lambda A(u),$$

for all $u, v \in G_C$ and $\lambda \in \mathbb{C}$.

We now consider an irreducible representation F of degree f. Each $c_\alpha (\alpha = 1, 2, \ldots, k)$ has the property that

$$c_\alpha x = x c_\alpha,$$

for all $x \in G$. On extending F to G_C we deduce that

$$F(c_\alpha)F(x) = F(x)F(c_\alpha). \quad (x \in G)$$

It follows from Schur's Lemma (p. 24) that

$$F(c_\alpha) = \lambda_\alpha I_f, \tag{5.17}$$

where λ_α is a complex number, which we proceed to determine. Let χ_α be the value of $\chi(x)$ when x lies in C_α. As usual, we put $h_\alpha = |C_\alpha|$. Hence on applying F to (5.15) and taking the trace we find that

$$\mathrm{tr}\, F(c_\alpha) = h_\alpha \chi_\alpha,$$

which in conjunction with (5.17) yields

$$\lambda_\alpha = \frac{h_\alpha \chi_\alpha}{f} \quad (\alpha = 1, 2, \ldots, k).$$

Next we operate with F on each term of (5.16). Using (5.17) and omitting the factor I_f throughout we obtain that

$$\lambda_\alpha \lambda_\beta = \sum_{\gamma=1}^{k} a_{\alpha\beta\gamma} \lambda_\gamma.$$

This is an instance of the situation discussed in the Lemma (p. 144), where we put $\xi = \lambda_\alpha$ for a fixed value of α and identify $y_1, y_2, \ldots y_N$ with $\lambda_1, \lambda_2, \ldots \lambda_k$ (note that $\lambda_1 = 1$). We have therefore established the following crucial fact:

Theorem 5.4. *Let χ be a simple character of degree f which takes the value χ_α for the conjugacy class C_α. Then each of the numbers $h_\alpha \chi_\alpha / f$ is an algebraic integer, where $h_\alpha = |C_\alpha|$.*

As a first application of this theorem we obtain valuable information about the degree of representations.

Theorem 5.5. *The degree of an absolutely irreducible representation is a factor of the order of the group.*

Proof. Let χ be a simple character of degree f. Then $\langle \chi, \chi \rangle = 1$, that is

$$\sum_{\alpha=1}^{k} h_\alpha \chi_\alpha \bar{\chi}_\alpha = g.$$

Hence

$$\frac{g}{f} = \sum_{\alpha=1}^{k} \frac{h_\alpha \chi_\alpha}{f} \bar{\chi}_\alpha.$$

Now, for each α, both $h_\alpha \chi_\alpha / f$ and $\bar{\chi}_\alpha$ are algebraic integers. Hence by Theorem 5.2, g/f is an algebraic integer and therefore a rational integer, that is f divides g.

5.3. Burnside's (p, q)-theorem

Among the most interesting general theorems about finite groups are those which deal with relationships between the structure of the group and arithmetical features of certain integers that are associated with the group in a natural manner. The Sylow theorems are typical examples of this kind: they assert, among other things, that if the order of G is divisible by p^a, where p is a prime, then G possesses a subgroup of order p^a. It is one of Burnside's finest achievements that he succeeded in exploiting the theory of group characters, then newly discovered, in order to establish arithmetical theorems of remarkable depth and elegance.

We begin with a somewhat technical result that forms the foundation of what follows.

148

Theorem 5.6. *Let F be an absolutely irreducible representation of degree f and character χ. Suppose that for a particular conjugacy class C_α, for which $h_\alpha = |C_\alpha|$, we have that $(h_\alpha, f) = 1$. Then either*
   (i) $\chi_\alpha = f\varepsilon_0$, *where ε_0 is a root of unity, or else*
   (ii) $\chi_\alpha = 0$.

Proof. Since $(h_\alpha, f) = 1$, we can find integers r and s such that

$$rh_\alpha + sf = 1.$$

Hence

$$r\frac{h_\alpha\chi_\alpha}{f} + s\chi_\alpha = \frac{\chi_\alpha}{f}.$$

By virtue of Theorem 5.4 and Proposition 5.3 each term on the left is an algebraic integer, whence

$$\lambda = \chi_\alpha/f$$

is also an algebraic integer. Let

$$x^n + a_1 x^{n-1} + \ldots + a_{n-1}x + a_n \tag{5.18}$$

be the minimal polynomial for λ and let

$$\lambda, \lambda', \ldots, \lambda^{(n-1)}$$

be the conjugates of λ, that is the set of roots of (5.18). Then

$$|\lambda\lambda' \ldots \lambda^{(n-1)}| = |a_n|, \tag{5.19}$$

where $|a_n|$ is a non-negative integer.

   We know that

$$\lambda = \chi_\alpha/f = (\varepsilon_1 + \varepsilon_2 + \ldots + \varepsilon_f)/f, \tag{5.20}$$

where $\varepsilon_1, \varepsilon_2, \ldots, \varepsilon_f$ are gth roots of unity. Hence (see p. 59) we have that

$$|\chi_\alpha/f| \leqslant (|\varepsilon_1| + |\varepsilon_2| + \ldots + |\varepsilon_f|)/f = 1.$$

Two cases arise:

   (i) $|\chi_\alpha/f| = 1$. Then $\varepsilon_1 = \varepsilon_2 = \ldots = \varepsilon_f = \varepsilon_0$, say, whence $\chi_\alpha = f\varepsilon_0$, which is the first alternative.
   (ii) $|\chi_\alpha/f| = |\lambda| < 1$. We claim that $a_n = 0$. For, if not, $|a_n|$ would be a positive integer and (5.19) would imply that

$$|\lambda\lambda' \ldots \lambda^{(n-1)}| \geqslant 1. \tag{5.21}$$

We shall show that (5.21) leads to a contradiction. To this end we must scrutinise the conjugates of λ (if any) in more detail. In (5.20) each ε_r is a power of

$$\varepsilon = \exp(2\pi i/g).$$

Thus there exist integers m_r such that

$$\varepsilon_r = \varepsilon^{m_r} \quad (r = 1, 2, \ldots, f).$$

Hence we may regard λ as a polynomial in ε, say

$$\lambda = T(\varepsilon) \tag{5.22}$$

with rational coefficients. By Proposition 5.2, the conjugates of λ are expressed analogously to (5.22), ε being replaced by one of its conjugates, ε^s, where s is a suitable integer (p. 143). For example

$$\lambda' = T(\varepsilon^s) = (\varepsilon_1^s + \varepsilon_2^s + \ldots + \varepsilon_f^s)/f,$$

whence $|\lambda'| \leq 1$. Similar inequalities hold for the other conjugates of λ. Since we are assuming that λ satisfies the strict inequality $|\lambda| < 1$, it would follow that

$$|\lambda\lambda' \ldots \lambda^{(n-1)}| < 1,$$

contrary to (5.21). Hence $a_n = 0$, as asserted.

Now the minimal polynomial (5.18) is irreducible over the field of rational numbers. In the present circumstances this can happen only if $n = 1$ and the minimal polynomial reduces to x. Its sole root is $\lambda = 0$. This disposes of the second alternative.

We now come to the first application to 'pure' group theory.

Theorem 5.7. *Suppose that the finite group G possesses a conjugacy class C_α with the property that*

$$h_\alpha = |C_\alpha| = p^t > 1,$$

where p is a prime. Then G has a non-trivial normal subgroup.

Proof. We use an indirect argument. Assume that G has the property mentioned in the theorem but that G is a *simple* group, that is, it has no non-trivial normal subgroups.

Let F be a non-trivial irreducible representation of G, having degree f and character χ. The mapping

$$x \to F(x)$$

is injective (faithful). For, if not, the kernel would be a non-trivial normal subgroup of G.

Next, we observe that there can be no element other than 1, such that $\chi(u) = \varepsilon_0 f$, where ε_0 is a root of unity. For this would imply (p. 59) that

$$F(u) = \varepsilon_0 I,$$

whence

$$F(x)F(u) = F(u)F(x)$$

$$F(xu) = F(ux)$$

for all x in G. Since F is injective, it would follow that

$$xu = ux,$$

so that u would be a central element. Thus G would have a non-trivial centre and therefore a non-trivial normal subgroup, contrary to our hypothesis. Hence the first alternative in Theorem 5.6 cannot occur in the present circumstances. Thus for every non-trivial simple character $\chi^{(r)}$ we have that either (i) $p \mid f^{(r)}$ or (ii) $p \nmid f^{(r)}$, whence $(f^{(r)}, h_\alpha) = 1$ for the particular α in question and consequently $\chi_\alpha^{(r)} = 0$.

We now consider the character $\rho(x)$ of the regular representation, for which, as we know, $\rho(1) = g$ and $\rho(x) = 0$ if $x \neq 1$. Since C_α is not the class of 1, we have that $\rho_\alpha = 0$. The Fourier analysis of ρ is (p. 45)

$$\rho(x) = \sum_{r=1}^{k} f^{(r)} \chi^{(r)}(x),$$

where $f^{(1)} = 1$ and $\chi^{(1)}(x) = 1$ for all x. When $x \in C_\alpha$, we obtain that

$$0 = 1 + \sum_{r=2}^{k} f^{(r)} \chi_\alpha^{(r)}. \tag{5.23}$$

As we have observed, either $\chi_\alpha^{(r)} = 0$ or $p \mid f^{(r)}$. Therefore, ignoring zero terms, we can write (5.23) in the form

$$0 = 1 + p\eta,$$

where η is a sum of character values and hence an algebraic integer. But this would lead to

$$\eta = -1/p,$$

which is absurd, because $-1/p$ is clearly not an algebraic integer. This completes the proof.

We now come to the celebrated theorem referred to in the heading of this section.

151

Theorem 5.8. *Let G be a group of order $p^a q^b$, where p and q are distinct primes, and a and b are positive integers. Then G is not simple.*

Proof. Let Q be a Sylow q-group of G, so that $|Q| = q^b$. Since Q is a prime-power group, its centre is non-trivial [**13**, p. 59]. Thus suppose that $v \neq 1$ and $v \in Z(Q)$, the centre of Q. Now either $v \in Z(G)$, in which case G has a non-trivial centre and, consequently, is not simple; or else $v \in C_\alpha$ for some α such that $h_\alpha = |C_\alpha| > 1$. Then (p. 72)

$$h_\alpha = [G:Z_v],$$

where Z_v is the centraliser of v in G. By the choice of v, Q is a subgroup of Z_v. Hence $|Z_v| = p^c q^b$, say, where $c \geq 0$. It follows that

$$h_\alpha = p^a q^b / p^c q^b = p^{a-c},$$

where $a - c > 0$. By virtue of Theorem 5.7, G cannot be simple.

Historical remark. Soon after the publication of Burnside's (p, q)-theorem the problem was posed to prove this result by purely group-theoretic methods, without recourse to character or representation theory. This challenge turned out to be exceedingly taxing and, after strenuous efforts over more than half a century, the matter was settled successfully by H. Bender [**1**].

Whatever one's attitude to insistence on 'purity of method', the ingenuity and perseverance of modern group theoreticians command respect and admiration. Nevertheless, Burnside's original proof deserves to retain its place in the text-book literature: it is a work of great mathematical beauty and presents an outstanding example of cross-fertilisation between different algebraic theories.

5.4. Frobenius groups

As we have already remarked (p. 60), character theory is well adapted to discovering normal subgroups of a given G; by Theorem 2.7, all we have to do is to examine the equation

$$\phi(x) = \phi(1), \tag{5.24}$$

where ϕ is an arbitrary character of G. One of the most powerful applications of this principle is due to Frobenius himself. The celebrated result which he proved in this way can either be couched in the language of permutation groups or else serves to describe an interesting class of abstract groups, which now bear his name. We choose here the concrete approach.

152

Theorem 5.9. *Let G be a transitive permutation group of degree n such that each permutation of G, other than the identity, leaves at most one of the objects fixed. Then those permutations which displace all the objects, together with the identity, form a normal subgroup of G of order n.*

Remark. We are supposing that, for each permutation in G, the number of fixed points takes one of the values n, unity or zero. As an example we cite the group S_3, in which the number of fixed points is three, unity or zero. The theorem states that the permutations

$$1, (123), (132)$$

constitute a normal subgroup of S_3.

Proof. We adhere to the terminology and notation introduced on p. 101. Thus we assume that G acts transitively on the objects $1, 2, \ldots, n$. The stabiliser of 1 is denoted by H or H_1, and there are permutations $p_1(=1)$, $p_2, \ldots, p_n$ in G which transform 1 into $1, 2, \ldots, n$ respectively. The stabiliser of i is the group

$$H_i = p_i^{-1} H p_i \qquad (i = 1, 2, \ldots, n)$$

and

$$G = Hp_1 \cup Hp_2 \cup \ldots \cup Hp_n \qquad (5.25)$$

is the coset decomposition of G with respect to H. We put

$$|G| = g, \qquad |H| = h,$$

whence by (5.25)

$$g = nh.$$

The hypothesis that no permutation of G, other than 1, has more than one fixed point is expressed by the formula

$$H_i \cap H_j = \{1\} \quad (i \neq j). \qquad (5.26)$$

The set of non-identity elements of any group K will be denoted by

$$K^{\#} = K \backslash \{1\}.$$

Then $H_i^{\#}$ is the set of permutations for which i is the sole fixed object. If W is the collection of permutations with no fixed objects we have the decomposition

$$G = \{1\} \cup H_1^{\#} \cup H_2^{\#} \cup \ldots \cup H_n^{\#} \cup W \qquad (5.27)$$

153

into disjoint subsets. It is obvious that 1 is the only permutation with n fixed objects. The permutations with precisely one fixed object are comprised in the union

$$H_1^{\#} \cup H_2^{\#} \cup \ldots \cup H_n^{\#}.$$

Since $|H_j^{\#}| = h - 1$, it follows that G has $n(h - 1)$ permutations with a single fixed object, and that there are

$$|W| = g - 1 - n(h - 1) = n - 1 \tag{5.28}$$

permutations which displace all the objects.

In a group with this structure it is easy to compute the induced character ψ^G, where ψ is an arbitrary character of $H (=H_1)$. We recall that

$$\psi^G(x) = \sum_{i=1}^{n} \psi(p_i x p_i^{-1})$$

with the usual convention that $\psi(t) = 0$ if $t \notin H$. Let $d \, (=\psi(1))$ be the degree of ψ. Then

$$\psi^G(1) = nd.$$

Next, let $u_j \in H_j$, $u_j \neq 1$, so that

$$u_j = p_j^{-1} u p_j, \tag{5.29}$$

where $u \in H$, $u \neq 1$. Then

$$\psi^G(u_j) = \sum_{i=1}^{n} \psi(p_i p_j^{-1} u p_j p_i^{-1}), \tag{5.30}$$

and the ith term in this sum can be non-zero only if

$$p_i p_j^{-1} u p_j p_i^{-1} \in H,$$

that is

$$p_j^{-1} u p_j \in p_i^{-1} H p_i (=H_i).$$

By (5.26) this is impossible unless $j = i$. Therefore (5.30) reduces to

$$\psi^G(p_j^{-1} u p_j) = \psi(u)$$

$(j = 1, 2, \ldots, n)$. Finally, if $w \in W$, then

$$\psi^G(w) = 0,$$

because $p_i w p_i^{-1} \notin H$ for all i. The values of ψ^G are displayed in Table 5.1 with the understanding that (5.29) holds.

Since we have precise information about the number of fixed objects under the action of G, it is easy to compute the character

$$\theta(x) = \nu(x) - 1,$$

where $\nu(x)$ is the number of objects left invariant by x.

Indeed, we need only consider representatives of the three types of permutation, namely 1, u_j and w, for which the number of fixed objects is n, unity and zero respectively. Hence

$$\theta(1) = n - 1, \qquad \theta(u_j) = 0, \qquad \theta(w) = -1.$$

	1	u_j	w
ψ^G	dn	$\psi(u)$	0
θ	$n-1$	0	-1
$\xi = \psi^G - d\theta$	d	$\psi(u)$	d
σ	h	0	h

Table 5.1.

The generalised character (§ 4.4)

$$\xi(x) = \psi^G(x) - d\theta(x) \tag{5.31}$$

has the remarkable property that

$$\xi(w) = \xi(1) \quad (w \in W), \tag{5.32}$$

as is evident from Table 5.1. It would appear that we have come close to achieving the aim expressed in (5.24). But two difficulties remain to be surmounted: only proper characters can be used to identify kernels, and elements other than those of W might satisfy (5.32). The first problem is solved by the following surprising discovery: if ψ is a simple character of H of degree d, then the function ξ defined in (5.31) is a simple character of G of degree d. Thus we are now assuming that $\psi(1) = d$ and $\langle \psi, \psi \rangle_H = 1$, whence

$$\sum_{u \in H^*} |\psi(u)|^2 = h - d^2. \tag{5.33}$$

The table shows that $\xi(1) = d > 0$. Hence it suffices to prove that

$$\sum_{x \in G} |\xi(x)|^2 = g.$$

155

Using (5.27) we have that

$$\sum_{x \in G} |\xi(x)|^2 = \xi(1)^2 + \sum_{j=1}^{n} \left(\sum_{u_j \in H_j^*} |\xi(u_j)|^2 \right) + \sum_{w \in W} |\xi(w)|^2.$$

Hence, by the table and by (5.28) and (5.33),

$$\sum_{x \in G} |\xi(x)|^2 = d^2 + n(h - d^2) + (n-1)d^2 = nh = g,$$

as required.

Turning to the second point, let

$$\psi^{(1)}, \psi^{(2)}, \ldots, \psi^{(l)}$$

be the complete set of simple characters of H and put

$$\psi^{(r)}(1) = d^{(r)} \quad (r = 1, 2, \ldots, l).$$

Let ρ be the character of the right-regular representation of H, that is, by (2.23) and (2.25),

$$\rho(1) = h, \quad \rho(u) = 0 \quad (u \neq 1),$$

$$\rho = \sum_{r=1}^{l} d^{(r)} \psi^{(r)}.$$

We construct the character

$$\sigma = \sum_{r=1}^{l} d^{(r)} \xi^{(r)} \tag{5.34}$$

of G, where

$$\xi^{(r)} = \psi^{(r)G} - d^{(r)} \theta \quad (r = 1, 2, \ldots, l).$$

The values of $\xi^{(r)}$ are analogous to those of ξ, where $\psi^{(r)}$ takes the place of ψ. Thus, with the notation (5.29),

$$\xi^{(r)}(1) = d^{(r)}, \qquad \xi^{(r)}(u_j) = \psi^{(r)}(u), \qquad \xi^{(r)}(w) = d^{(r)}.$$

Substituting in (5.34) we obtain that

$$\sigma(1) = \sigma(w) = \sum_{r=1}^{l} (d^{(r)})^2 = h,$$

$$\sigma(u_j) = \sum_{r=1}^{l} d^{(r)} \psi^{(r)}(u) = \rho(u) = 0.$$

This makes it plain that σ is the character of a representation of G whose kernel is

$$1 \cup W,$$

which is therefore a normal subgroup of G of order n. This completes the proof.

Finally, we recast the result in an abstract form.

Theorem 5.10. *Suppose that H is a subgroup of a finite group G with the property that*

$$y^{-1}Hy \cap H = \{1\}, \tag{5.35}$$

whenever $y \in G$ but $y \notin H$. Then G has a normal subgroup K such that $G = HK$ and $H \cap K = \{1\}$.

Proof. Let

$$G = Ht_1 \cup Ht_2 \cup \ldots \cup Ht_n \quad (t_1 = 1)$$

be the coset decomposition of G with respect to H and consider the permutation representation

$$\pi(x) = \begin{pmatrix} Ht_1 & Ht_2 & \ldots & Ht_n \\ Ht_1x & Ht_2x & \ldots & Ht_nx \end{pmatrix} \tag{5.36}$$

of G. The coset Ht_i remains fixed under the action of $\pi(x)$ if $Ht_ix = Ht_i$, that is $t_ixt_i^{-1} \in H$, or

$$x \in t_i^{-1}Ht_i.$$

Suppose that $x \neq 1$. Then $\pi(x)$ cannot have a second fixed element; for this would imply that

$$x \in t_j^{-1}Ht_j \quad (i \neq j),$$

whence

$$t_i^{-1}Ht_i \cap t_j^{-1}Ht_j \neq \{1\},$$

$$H \cap t_it_j^{-1}Ht_jt_i^{-1} \neq \{1\},$$

which would contradict (5.35), because $t_jt_i^{-1} \notin H$. In particular, $\pi(x)$ is not the identity permutation unless $x = 1$. Thus the homomorphism

$$x \to \pi(x)$$

is injective. Hence G is faithfully represented by the permutations (5.36), which form a permutation group isomorphic to G. For simplicity we

157

shall identify these two groups, that is we shall regard x either as an element of the abstract group G or else as the permutation specified in (5.36). We can now apply Theorem 5.9 to G, viewed as a permutation group, because no element, other than 1, leaves more than one coset fixed. If W is the set of elements which displace all the cosets, then the theorem tells us that

$$K = 1 \cup W$$

is a normal subgroup of order n. We claim that the elements of K are in one-to-one correspondence with the cosets of H. Since every element of G lies in some coset, it suffices to show that no coset contains two distinct elements of K. Suppose, on the contrary, that w and w' belong to $Ht_i \cap K$. Then $w'w^{-1}$ belongs to H and hence fixes the coset Ht_1 $(=H)$; also $w'w^{-1}$ lies in K, because K is a group. But the identity is the only element of K that leaves anything fixed. It follows that $w' = w$, and this proves the assertion. We may therefore label the elements of K as

$$w_1 \, (=1), w_2, \ldots, w_n$$

in such a way that

$$w_i \in Ht_i \quad (i = 1, 2, \ldots, n),$$

and w_i may be used as a coset representative in place of t_i. It follows that $G = HK$. The relationship

$$H \cap K = \{1\}$$

has already been noted. This completes the proof.

Groups with the structure described in this theorem are called *Frobenius groups*.

Exercises

1. Let G be a group of order pq, where p and q are primes such that $p < q$. Show that if $|G'| = q$, the number of conjugacy classes is $p + (q-1)p^{-1}$.

2. In the notation of p. 146 let

$$c_\alpha c_\beta = \sum_{\gamma=1}^{k} a_{\alpha\beta\gamma} c_\gamma$$

be the structure equations for the centre of the group algebra G_c. Show that

$$a_{\alpha\beta\gamma} = \frac{h_\alpha h_\beta}{g} \sum_{j=1}^{k} \chi_\alpha^{(j)} \chi_\beta^{(j)} \bar{\chi}_\gamma^{(j)} / f^{(j)}.$$

Prove that the elements

$$e^{(i)} = (f^{(i)}/g) \sum_\alpha \bar{\chi}_\alpha^{(i)} c_\alpha \quad (i = 1, 2, \ldots, k)$$

of G_C satisfy the relations

$$e^{(i)} e^{(j)} = \delta_{ij} e^{(i)}.$$

(Thus G_C possesses k mutually orthogonal central idempotents.)

3. Prove that if the order of G is equal to p^4, where p is a prime, then the degree of an absolutely irreducible representation of G is either equal to unity or else equal to p.

4. Prove that, for a group of odd order g and class number k, the integer $g - k$ is divisible by 16.

6

ARITHMETIC PROPERTIES OF GROUP CHARACTERS

6.1. Real character values

The question of real-valued characters was briefly considered in Exercises 4 to 6 of Chapter 2.

In this section we are concerned with a particular element g of a finite group G and an arbitrary character ψ, which need not be irreducible. We have seen in (2.87) (p. 51) that

$$\psi(g^{-1}) = \overline{\psi(g)}. \tag{6.1}$$

Hence we have

Proposition 6.1. *The character value $\psi(g)$ is real if and only if*

$$\psi(g^{-1}) = \psi(g). \tag{6.2}$$

We recall that all characters are class functions (Proposition 1.1(ii)): thus

$$\psi(x) = \psi(y)$$

if x and y are conjugate in G, which we write as

$$x \sim y.$$

Hence if $g^{-1} \sim g$, then equation (6.2) holds for every character ψ, and $\psi(g)$ is therefore real. We shall now show that the converse is also true.

Theorem 6.1. *The numbers $\psi(g)$ are real for all characters ψ of a finite group if and only if*

$$g^{-1} \sim g. \tag{6.3}$$

Proof. It only remains to prove that (6.3) is a necessary condition for the reality of all $\psi(g)$. Suppose that g^{-1} and g belong to distinct conjugacy classes C_α and C_β respectively. Put

$$\psi_\alpha = \psi(g^{-1}), \quad \psi_\beta = \psi(g).$$

Then by (6.1)

$$\psi_\alpha = \overline{\psi}_\beta.$$

160

In particular, for each irreducible character $\chi^{(i)}$ we have that

$$\chi_\alpha^{(i)} = \overline{\chi_\beta^{(i)}}.$$

In this case the character relations of the second kind ((2.42), p. 51) imply that

$$\sum_{i=1}^{k} (\chi_\alpha^{(i)})^2 = 0.$$

It is therefore evident that not all the values of $\chi_\alpha^{(i)}$ can be real if (6.3) is false.

Example 1. In the group A_4 (Table 2.5(b), p. 61) the elements (123) and $(123)^{-1}$ $(=(132))$ are not conjugate; they belong to the classes C_2 and C_3 respectively. Hence not all characters of A_4 are real.

Example 2. In the group A_5 (Table 3.1, p. 79) each element is conjugate to its inverse. For an element that is conjugate to (12)(34) satisfies $g^2 = 1$ and so $g = g^{-1}$; the class of (123) contains all 3-cycles and in particular $(123)^{-1}$; if $g = (12345)$, then $g \sim g^4 (= g^{-1})$ and $g^2 \sim (g^2)^4 = g^{-2}$.

6.2. Rational character values

We have seen (Proposition 5.3, p. 146) that all character values are algebraic integers. If a particular character value $\psi(g)$ is also rational, then it must be a rational, that is, an ordinary, integer. In this section we shall obtain conditions for this situation to occur.

An algebraic number is rational if and only if its minimal polynomial over $\mathbb{Q}$ (p. 141) is of degree unity, which means that it has no algebraic conjugate other than itself.

We begin by elaborating a remark made at the end of § 5.1. Let h be a positive integer and put

$$\varepsilon = \exp(2\pi i/h). \tag{6.4}$$

The number of integers r that satisfy

$$(h, r) = 1, \quad 1 \leqslant r \leqslant h \tag{6.5}$$

is equal to the Euler function

$$\phi(h) = h \prod_{p|h} \left(1 - \frac{1}{p}\right), \tag{6.6}$$

where p ranges over all prime divisors of h [11, p. 52].

161

The algebraical conjugates of ε are described by the following theorem, which we quote without proof (see P. M. Cohn, *Algebra*. Wiley & Sons, 1977, Vol. 2, § 5.6).

Theorem 6.2. Let $\varepsilon = \exp(2\pi i/h)$. The conjugates of ε are the numbers ε^r, where $(h, r) = 1$. The minimal polynomial for ε over $\mathbb{Q}$ is of degree $\phi(h)$. It is called the cyclotomic polynomial of order h and is given by

$$\Phi_h(t) = \prod_{(r, h) = 1} (t - \varepsilon^r).$$

Example 3. When $h = 12$, we have that

$$\varepsilon = \exp(\pi i/6) = \tfrac{1}{2}\sqrt{3} + \tfrac{1}{2}i, \quad \varepsilon^2 = \exp(\pi i/3) = \tfrac{1}{2} + \tfrac{1}{2}\varepsilon\sqrt{\,}\varepsilon,$$

and

$$r = 1, 5, 7, 11.$$

Hence

$$\Phi_{12}(t) = (t - \varepsilon)(t - \varepsilon^5)(t - \varepsilon^7)(t - \varepsilon^{11})$$

$$= (t - \varepsilon)(t - \varepsilon^5)(t + \varepsilon)(t + \varepsilon^5)$$

$$= (t^2 - \varepsilon^2)(t^2 - \varepsilon^{10}) = (t^2 - \varepsilon^2)(t^2 - \varepsilon^{-2}),$$

so
so

$$\Phi_{12}(t) = t^4 - t^2 + 1.$$

Example 4. If p is a prime

$$\Phi_p(t) = \frac{t^p - 1}{t - 1} = t^{p-1} + t^{p-2} + \cdots + t + 1 \tag{6.8}$$

and

$$\Phi_{p^\alpha}(t) = \frac{t^{p^\alpha} - 1}{t^{p^{\alpha-1}} - 1} = t^{p^{\alpha-1}(p-1)} + t^{p^{\alpha-1}(p-2)} + \cdots + 1 \tag{6.9}$$

$$= \Phi_p(t^{p^{\alpha-1}}),$$

where α is a positive integer.

The question of the rationality of character values is now easily settled:

Theorem 6.3. Let g be an element and let ψ be a character of a finite group G. Suppose that g is of order h. Then $\psi(g)$ is rational if

$$\psi(g^r) = \psi(g) \tag{6.10}$$

for all integers r satisfying (6.5). Moreover, $\psi(g)$ is rational for all ψ if and only if

$$g^r \sim g \qquad (6.11)$$

for all r satisfying (6.5).

Proof. Let $A(x)$ be the representation of G of degree m which affords the character $\psi(x)$. Since g is a particular element, we may choose the basis of the underlying G-module in such a way that

$$A(g) = \text{diag}(\varepsilon_1, \varepsilon_2, \ldots, \varepsilon_m). \qquad (6.12)$$

Hence, for every integer k,

$$A(g^k) = \text{diag}(\varepsilon_1^k, \varepsilon_2^k, \ldots, \varepsilon_m^k). \qquad (6.13)$$

Each latent root ε_j satisfies the equation

$$t^h - 1 = 0$$

and is therefore a power of ε, as defined in (6.4). Hence

$$\psi(g) = \varepsilon_1 + \varepsilon_2 + \ldots + \varepsilon_m = d_0 + d_1 \varepsilon + d_2 \varepsilon^2 + \ldots + d_{h-1} \varepsilon^{h-1},$$

where $d_s (\geqslant 0)$ $(s = 0, 1, \ldots, h-1)$ denotes the number of times that ε^s occurs in (6.12). We use the notation

$$\psi(g) = T(\varepsilon),$$

where

$$\qquad (6.14)$$

$$T(t) = d_0 + d_1 t + \ldots + d_{h-1} t^{h-1}$$

is a polynomial in $\mathbb{Q}[t]$. According to Proposition 5.2 (p. 143) and Theorem 6.2, the conjugates of $T(\varepsilon)$ are of form $T(\varepsilon^r)$ where r satisfies (6.5). Now an algebraic number is rational if and only if it has no conjugates other than itself, which means that

$$T(\varepsilon^r) = T(\varepsilon) \qquad (6.15)$$

for all r in (6.5). By (6.13) and (6.14) the condition (6.15) is equivalent to (6.10), thus proving the first assertion of the theorem. Next we observe that (6.11) implies that (6.10) holds for every ψ so that all character values $\psi(g)$ are rational for this element g. Conversely, we shall show that if (6.10) holds for all characters ψ then (6.11) must be true: if not, assume that, for some r_0, the elements g^{r_0} and g belong to distinct conjugacy classes C_α and C_β respectively. In particular for each irreducible character $\chi^{(i)}$ the hypothesis

163

(6.10) implies that

$$\chi^{(i)}(g^{r_0}) = \chi^{(i)}(g);$$

so

$$\chi_\alpha^{(i)} = \chi^{(i)}(g^{r_0}), \quad \chi_\beta^{(i)} = \chi^{(i)}(g)$$

and it would follow that

$$\chi_\alpha^{(i)} = \chi_\beta^{(i)} = \bar{\chi}_\beta^{(i)},$$

because all character values for g are real. However, this is incompatible with (2.42) (p. 51) which in this case reduces to

$$\sum_{i=1}^{k} (\chi_\alpha^{(i)})^2 = 0.$$

The most important application of this theorem is described in the following result (see p. 123).

Proposition 6.2. *For each n ($\geqslant 1$) the symmetric group S_n has the property that all character values are rational integers.*

Proof. Let $c = (\alpha_1, \alpha_2, \ldots, \alpha_l)$ be a cycle of order l in S_n. It is plain that when a power of c is decomposed into a product of disjoint cycles it cannot involve cycles of order greater than l, though shorter cycles may occur. For example,

$$(1\ 2\ 3\ 4)^3 = (1\ 4\ 3\ 2),$$

$$(1\ 2\ 3\ 4)^2 = (1\ 3)(2\ 4).$$

Let

$$g = c_1 c_2 \ldots c_t \tag{6.16}$$

be an arbitrary element of S_n, decomposed into disjoint cycles. We use the notation

$$\mu(g) = length\ of\ the\ longest\ cycle\ in\ g. \tag{6.17}$$

Then by the above observation

$$\mu(c^r) \leqslant \mu(c) \quad (r = 0, 1, 2, \ldots)$$

and by applying this result to each factor in (6.16)

$$\mu(g^r) \leqslant \mu(g). \tag{6.18}$$

If, in addition,

$$(r, l) = 1, \tag{6.19}$$

164

we can find integers u and v such that

$$1 = ru + lv.$$

Hence

$$c = (c^r)^u (c^l)^v = (c^r)^u,$$

that is c is a power of c^r. It follows from (6.18) that

$$\mu(c) \leqslant \mu(c^r).$$

Therefore (6.19) implies that

$$\mu(c) = \mu(c^r). \tag{6.20}$$

Suppose that g is of order h. Then

$$h = \text{l.c.m.}(l_1, l_2, \ldots, l_t),$$

where l_j is the order of c_j and l.c.m. stands for 'least common multiple'. We now assume that

$$(r, h) = 1;$$

then

$$(r, l_j) = 1 \quad (j = 1, 2, \ldots, t)$$

and it follows that the cycle decomposition of g^r is given by

$$g^r = c_1^r c_2^r, \ldots, c_t^r,$$

where c_j^r is a cycle of order l_j by virtue of (6.20). Thus g^r and g have the same cycle pattern and are therefore conjugate in S_n [13, Proposition 21, p. 131]. Hence, by Theorem 6.3, all character values of S_n are rational integers.

6.3. A congruence property

There are some useful congruence relations between character values. For the sake of simplicity we shall confine ourselves in this section to the case of rational character values. A more general situation will be discussed in §A.5 of the Appendix (p. 197).

Proposition 6.3. *Let g be an element of a finite group G and suppose that g is of order p^a, where p is a prime and a is a positive integer. Let ψ be a character of degree m ($= \psi(1)$) such that $\psi(g)$ is a rational integer. Then*

$$\psi(g) \equiv m \pmod{p}. \tag{6.21}$$

(I am indebted to G. D. James for drawing my attention to these results and for communicating Example 5 (p. 166).)

Proof. Let $\varepsilon = \exp(2\pi i/p^a)$. The minimal equation of ε over $\mathbb{Q}$ is given in (6.9). Since there are precisely p terms on the right-hand side of that equation we have that

$$\Phi_{p^a}(1) = p. \tag{6.22}$$

Put

$$\psi(g) = b, \tag{6.23}$$

where, by hypothesis, b is a rational integer. Using the notation of (6.14) we have that

$$T(\varepsilon) - b = 0.$$

This is an equation for ε with rational, and indeed integral, coefficients. As we have seen on p. 142, it follows that the polynomial $T(t) - b$ is divisible by the minimal polynomial for ε; thus

$$T(t) - b = \Phi_{p^a}(t)Q(t). \tag{6.24}$$

Since the leading coefficient of Φ_{p^a} is equal to unity, the division of $T(t) - b$ by $\Phi_{p^a}(t)$ yields a quotient $Q(t)$ whose coefficients are rational integers. In particular,

$$Q(1) = w \tag{6.25}$$

say, where w is an integer. Referring to (6.14) we note that

$$T(1) = d_0 + d_1 + \ldots + d_{h-1}.$$

The right-hand side of this equation enumerates all the latent roots of $A(g)$, and so

$$T(1) = m. \tag{6.26}$$

On putting $t = 1$ in (6.24) and using (6.22), (6.25) and (6.26) we obtain that

$$m - b = pw,$$

whence

$$m \equiv b \pmod{p},$$

which, by (6.23), proves the assertion.

Example 5. Although the complete character table of the group S_5 was obtained on p. 135, let us suppose, for the purpose of this example, that we possess only the following information about this group:

(1) the generators and sizes of the seven conjugacy classes as shown in Table 4.2 (p. 136);

(2) the fact that all character values are rational integers (Proposition 6.2);

(3) the values of the trivial and alternating characters.

We know that there are seven irreducible characters. In contrast to Table 4.2, these will be denoted by $\chi^{(i)}$ ($i = 1, 2, \ldots, 7$), the precise ordering being determined later, except that $\chi^{(1)}$ and $\chi^{(2)}$ will refer to the trivial and alternating characters respectively.

Let $u = (1\,2\,3\,4\,5)$, which generates the class (5). Applying equation (2.42) to the case in which $\alpha = \beta = (5)$ we obtain that

$$\sum_{i=1}^{7} (\chi^{(i)}(u))^2 = \tfrac{120}{4} = 5. \tag{6.27}$$

Since $\chi^{(i)}(u)$ is a rational integer, it follows that on the left-hand side of (6.27) precisely five terms are equal to unity and the remaining two terms are equal to zero, because we know that

$$\chi^{(1)}(u) = \chi^{(2)}(u) = 1. \tag{6.28}$$

With suitable numbering of the other characters the values of $\chi^{(i)}(u)$ may be provisionally listed as follows:

	$\chi^{(1)}$	$\chi^{(2)}$	$\chi^{(3)}$	$\chi^{(4)}$	$\chi^{(5)}$	$\chi^{(6)}$	$\chi^{(7)}$
u	1	1	± 1	± 1	± 1	0	0

Table 6.1.

We now turn our attention to the degrees $f^{(i)}$ ($= \chi^{(i)}(1)$) of the irreducible representations. By (2.26)

$$\sum_{i=1}^{7} (f^{(i)})^2 = 120. \tag{6.29}$$

On applying (6.21) to $\chi^{(6)}$ and $\chi^{(7)}$ when $g = u$, and hence $p = 5$, we find that

$$f^{(6)} = 5 \text{ or } 10, \ldots; \quad f^{(7)} = 5 \text{ or } 10 \ldots.$$

Evidently the only values that are compatible with (6.29) are

$$f^{(6)} = f^{(7)} = 5. \tag{6.30}$$

Equation (6.29) now reduces to

$$(f^{(3)})^2 + (f^{(4)})^2 + (f^{(5)})^2 = 120 - 1 - 1 - 25 - 25 = 68. \tag{6.30}$$

Again, when $g = u$, (6.21) implies that each of $f^{(3)}$, $f^{(4)}$ or $f^{(5)}$ is either equal to 4 or equal to 6. But the only combination of these values that satisfies (6.30) is

$$4^2 + 4^2 + 6^2 = 68.$$

Hence with suitable numbering of the characters we may put

$$f^{(3)} = f^{(4)} = 4, \quad f^{(5)} = 6.$$

Thus we have obtained the degrees of all the irreducible representations of S_5, which we summarise in the following table:

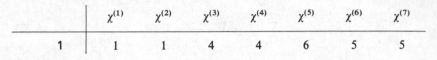

	$\chi^{(1)}$	$\chi^{(2)}$	$\chi^{(3)}$	$\chi^{(4)}$	$\chi^{(5)}$	$\chi^{(6)}$	$\chi^{(7)}$
1	1	1	4	4	6	5	5

Table 6.2.

By virtue of the character relations of the second kind the entries in the Tables 6.1 and 6.2 are orthogonal, that is

$$1 + 1 \pm 4 \pm 4 \pm 6 = 0.$$

It is easy to see that this equation can hold only if the minus sign is attached to the third and fourth terms and the plus sign to the last term. Hence, finally, the character values for u are as follows:

	$\chi^{(1)}$	$\chi^{(2)}$	$\chi^{(3)}$	$\chi^{(4)}$	$\chi^{(5)}$	$\chi^{(6)}$	$\chi^{(7)}$
u	1	1	-1	-1	1	0	0

Table 6.3.

Exercises

1. Show that in the dihedral group

$$D_n = gp\{a, b \,|\, a^n = b^2 = 1, bab = a^{-1}\}$$

every element is conjugate to its inverse. Deduce that all characters of D_n are real-valued (see Exercise 12 of Chapter 2, p. 66).

2. The dicyclic group of order $4n$ ($n > 1$) is given by

$$\Delta_n = gp\{a, b \,|\, a^{2n} = 1, a^n = (ab)^2 = b^2\}.$$

Prove that, when n is odd, not all the characters of Δ_n are real-valued (see Exercise 9 of Chapter 2, p. 65).

168

3. Show that, in the group S_n, a full cycle

$$z = (1 \; 2 \ldots n)$$

commutes only with elements z^k $(k = 0, 1, \ldots, n-1)$. (See [13], Chapter 7, Exercise 5.) Prove that, when $n = 4m - 1$ $(m \geqslant 1)$, the equation

$$t^{-1} z t = z^{-1}$$

implies that t is odd. Deduce that not all the characters of A_n are real-valued.

4. Construct the complete character table of S_4 by using only
   (a) the information given in Table 6.4 below;
   (b) the orthogonality relations for simple characters;
   (c) the fact that all character values are rational integers.

	(1)	(12)	(123)	(12)(34)	(1234)
h	1	6	8	3	6
$\chi^{(1)}$	1	1	1	1	1
$\chi^{(2)}$	1	-1	1	1	-1

Table 6.4.

(The remaining characters $\chi^{(3)}$, $\chi^{(4)}$, $\chi^{(5)}$ may be arranged conveniently.)

7

REAL REPRESENTATIONS

7.1. Statement of the problem

Up to now it has been taken for granted that the coefficients of an irreducible representation

$$F(x) \quad (x \in G) \tag{7.1}$$

are, in general, complex numbers. However, as we do not regard equivalent representations as distinct, we are led to the following:

Question. Even though $F(x)$ is complex for some x, does there exist an invertible matrix T such that

$$T^{-1}F(x)T \tag{7.2}$$

is real for all x?

In order to avoid circumlocution we introduce the following conventions.

Terminology. Let Φ be a scalar or matrix function defined on G. We say

(1) Φ is real if $\Phi(x)$ is real for all x in G;
(2) Φ is complex if $\Phi(x)$ is not real for at least one element x;
(3) Φ is (complex) when it is left open whether Φ is real or complex.

Unless the contrary is stated, F denotes an irreducible representation with simple character χ.

Clearly, if F is equivalent to a real representation, then χ must be real; for (7.1) and (7.2) have the same character and the latter is real. But the converse is not true; it may happen that χ is real, but F is not equivalent to a real representation. In any event, if the matrices $F(x)$ form an irreducible representation of G, so do the conjugate complex matrices

$$\bar{F}(x) \quad (x \in G) \tag{7.3}$$

because

$$\bar{F}(x)\bar{F}(y) = \bar{F}(xy).$$

170

The character of $\bar{F}$ is $\bar{\chi}$. Hence χ is real ($\chi = \bar{\chi}$) if and only if F and $\bar{F}$ are equivalent, though they need not be identical (that is real). Therefore, we distinguish three cases:

(I) χ is real and F is equivalent to a real representation.

(II) χ is real, but F is not equivalent to a real representation.

(III) χ is complex; then F and $\bar{F}$ are inequivalent and neither of them is equivalent to a real representation.

The object of this chapter is to deriva a formula, expressed in terms of χ only, which discriminates between the three cases. The result is due to G. Frobenius and I. Schur whose original memoir [9b] we shall follow.

7.2. Quadratic forms

Let
$$x_1, x_2, \ldots, x_n$$

be a set of commuting variables (indeterminates). A quadratic form is a polynomial of the form

$$q = a_{11}x_1^2 + \ldots + a_{nn}x_n^2 + 2(a_{12}x_1x_2 + \ldots + a_{n-1,n}x_{n-1}x_n) \qquad (7.4)$$

involving $\frac{1}{2}n(n+1)$ coefficients a_{ij} ($1 \leqslant i \leqslant j \leqslant n$). In many applications the coefficients are assumed to be real; but we shall here consider the more general case in which they may be arbitrary complex numbers.

A much more convenient way to express (7.4) is as follows: introduce the row vector of variables

$$\mathbf{x} = (x_1, x_2, \ldots, x_n)$$

and the symmetric matrix
$$A = (a_{ij}) \quad (a_{ji} = a_{ij}),$$

that is
$$A' = A.$$

Then
$$q = \mathbf{x}A\mathbf{x}', \qquad (7.5)$$

the factor 2 in (7.4) being accounted for by the fact that

$$a_{ij}x_ix_j = a_{ji}x_ix_j \quad (i \neq j).$$

Consider a linear transformation of the variables, thus

$$\mathbf{z} = \mathbf{x}P, \qquad (7.6)$$

where

$$\mathbf{z} = (z_1, z_2, \ldots, z_n)$$

is a set of variables and P is an invertible matrix. The quadratic form then becomes

$$q = \mathbf{z}P^{-1}A(P^{-1})'\mathbf{z}' = \mathbf{z}B\mathbf{z}'$$

say, where

$$B = P^{-1}A(P^{-1})'. \tag{7.7}$$

The reader will be familiar with the fundamental fact that a non-zero real quadratic form can be reduced to a sum or difference of squares by means of a suitable linear transformation. An analogous result holds in the complex field, except that we can now dispense with 'negative' squares:

Theorem 7.1. *Let* $q = \mathbf{x}A\mathbf{x}'$ *be a non-zero quadratic form, where A is a (complex) symmetric matrix. Then there exists an integer r satisfying*

$$1 \leqslant r \leqslant n$$

and an invertible matrix P such that

$$q = z_1^2 + z_2^2 + \ldots + z_r^2,$$

where

$$\mathbf{x} = \mathbf{z}P.$$

When $r = n$, we have that

$$A = PP'. \tag{7.8}$$

The proof, which is similar to that used for real quadratic forms, is given in §A.4 of the Appendix (p. 194).

As the reader will know, the 'pure sum' theorem holds for an important class of real quadratic forms. It is convenient to recall here the relevant definitions and results:

Definition 7.1. *The real quadratic form*

$$q = \mathbf{x}A\mathbf{x}'$$

and the corresponding real symmetric matrix A are said to be positive definite if, for every non-zero vector $\mathbf{u}$, we have that

$$\mathbf{u}A\mathbf{u}' > 0. \tag{7.9}$$

We note the following properties of positive definite matrices:

(1) Using the vector

$$\mathbf{e}_k = (0, \ldots, 1, \ldots, 0),$$

which has unity as its kth component and zeros elsewhere, we obtain that

$$\mathbf{e}_k A \mathbf{e}_k' = a_{kk} > 0. \tag{7.10}$$

Thus in a positive definite matrix all diagonal elements are positive.

(2) Every positive definite real matrix is invertible. For, if not, we could find a non-zero vector $\mathbf{u}$ such that

$$\mathbf{u}A = \mathbf{0},$$

whence

$$\mathbf{u}A\mathbf{u}' = 0$$

in contradiction to (7.9).

(3) We state the analogue of (7.8):

Proposition 7.1. *Let A be a positive definite real matrix. Then there exists a real invertible matrix P such that*

$$A = PP'.$$

A short, though somewhat inappropriate, proof is indicated in Exercise 1 at the end of this chapter.

7.3. Orthogonal representations

We recall that a square matrix is said to be *orthogonal* if

$$RR' = R'R = I.$$

In most contexts, R is assumed to be real; but the concept of an orthogonal matrix is valid in the complex field, and we shall use it here in this wider sense.

Definition 7.2. *A (complex) representation $R(x)$ of G is called orthogonal if, for each x in G, we have that*

$$R(x)R'(x) = R'(x)R(x) = I. \tag{7.11}$$

Next we shall use an idea which was already alluded to in Exercise 20 of Chapter 2 (p. 68). It is convenient to repeat briefly the relevant definitions and facts.

(1) An $n \times n$ matrix H is called *Hermitian* if

$$\bar{H}' = H. \tag{7.12}$$

173

(2) If H is a Hermitian matrix and if $\mathbf{u}$ is a (complex) row vector, then

$$h(\mathbf{u}) = \mathbf{u} H \mathbf{u}' \tag{7.13}$$

is a real number; and H is said to be positive definite if

$$h(\mathbf{u}) > 0$$

whenever $\mathbf{u} \neq \mathbf{0}$.

(3) Every positive definite Hermitian matrix is invertible.

(4) If H is positive definite, so are

$$\bar{H} \; (=H') \quad \text{and} \quad H^{-1}.$$

Definition 7.3. *The Hermitian matrix K is said to be an Hermitian invariant of the representation $A(x)$ $(x \in G)$ if*

(i) *K is positive definite, and*

(ii) *$A(x) K \bar{A}'(x) = K$ for all x in G.* $\tag{7.14}$

Proposition 7.2. *Every representation $A(x)$ of a finite group G possesses Hermitian invariants.*

Proof. Let

$$K = \sum_{y \in G} A(y) \bar{A}'(y). \tag{7.15}$$

It is easily verified that K is indeed an Hermitian invariant of $A(x)$ (see Exercise 20 of Chapter 2).

We note that if K is an Hermitian invariant of $A(x)$, so is

$$H = aK, \tag{7.16}$$

where a is an arbitrary positive constant.

By means of an ingenious use of matrix algebra, Frobenius and Schur established the following remarkable result.

Theorem 7.2. *Every (complex) irreducible orthogonal representation of a finite group G is equivalent to a real orthogonal representation.*

Proof. 1. As in (7.15) we use the Hermitian invariant

$$K = \sum_{y \in G} R(y) \bar{R}'(y)$$

174

so that

$$R(x)K\bar{R}'(x)=K. \tag{7.17}$$

On taking the complex conjugate of (7.11) we obtain that

$$\bar{R}(x)\bar{R}'(x)=\bar{R}'(x)\bar{R}(x)=I. \tag{7.18}$$

Hence (7.17) can be written as

$$R(x)K=K\bar{R}(x). \tag{7.19}$$

Transposing this equation we have that

$$K'R'(x)=\bar{R}'(x)K',$$

or, equivalently, by (7.11) and (7.18),

$$\bar{R}(x)K'=K'R(x). \tag{7.20}$$

Using (7.19) and (7.20) we find that

$$R(x)(KK')=(R(x)K)K'=K(\bar{R}(x)K')$$

$$=(KK')R(x).$$

By the Corollary to Schur's Lemma (p. 27) it follows that

$$KK'=\alpha I, \tag{7.21}$$

where α is a number. We shall now show that α is, in fact, real and positive. For (7.21) implies that

$$KK'K=\alpha K.$$

Hence if **u** is an arbitrary non-zero vector we obtain that

$$(\mathbf{u}K)K'(K\bar{\mathbf{u}}')=\alpha(\mathbf{u}K\bar{\mathbf{u}}'). \tag{7.22}$$

Let

$$\mathbf{v}=\mathbf{u}K.$$

Then (7.22) becomes

$$\mathbf{v}K'\bar{\mathbf{v}}'=\alpha(\mathbf{u}K\bar{\mathbf{u}}');$$

it is now obvious that $\alpha>0$, because both K and K' are positive definite and $\mathbf{v}\neq\mathbf{0}$.

In accordance with (7.16) we may replace K by the Hermitian invariant H given by

$$H=(\alpha)^{-1/2}K. \tag{7.23}$$

It is convenient to summarise the properties of H:

175

(i) H is positive definite Hermitian, that is

$$\bar{H}' = H \tag{7.24}$$

and $\mathbf{u}H\bar{\mathbf{u}}' > 0$ if $\mathbf{u} \neq \mathbf{0}$.

(ii) $R(x)H\bar{R}'(x) = H \quad (x \in G)$. $\tag{7.25}$

(iii) $HH' = I = H'H$. $\tag{7.26}$

This last property is established by substituting (7.23) in (7.21). Thus H is both Hermitian and orthogonal.

2. Since H is positive definite, so is $I + H'$. Hence $I + H'$ is invertible and we can construct the representation

$$S(x) = (I + H')R(x)(I + H')^{-1}, \tag{7.27}$$

which is equivalent to $R(x)$. The crucial step in the proof consists in demonstrating that $S(x)$ is real. Taking conjugates in (7.27) and using (7.24) we find that

$$\bar{S}(x) = (I + H)\bar{R}(x)(I + H)^{-1}.$$

Next, (7.25) can be rewritten as

$$\bar{R}(x) = H'R(x)H = H'R(x)H'^{-1},$$

by (7.18). Hence

$$\bar{S}(x) = (I + H)H'R(x)\{(I + H)H'\}^{-1},$$

$$\bar{S}(x) = (H' + I)R(x)(H' + I)^{-1}$$

by (7.26). Thus

$$\bar{S}(x) = S(x),$$

which proves that $S(x)$ is real.

3. Although $S(x)$ is real and equivalent to $R(x)$, it is not an orthogonal representation. In fact, it is readily verified that

$$S(x)QS'(x) = Q, \tag{7.28}$$

where

$$Q = (I + H')(I + H). \tag{7.29}$$

Using (7.24) and (7.26) we can write

$$Q = 2I + H + H' = 2I + H + \bar{H}.$$

This formula shows that Q is real, symmetric and positive definite, because I, H and $\bar{H}$ are positive definite. Hence, by Proposition 7.1, there exists an

176

invertible real matrix P such that

$$Q = PP',$$

and (7.28) becomes

$$\{P^{-1}S(x)P\}\{P^{-1}S(x)P\}' = I.$$

Thus the representation

$$T(x) = P^{-1}S(x)P,$$

which is equivalent to $S(x)$ and hence to $R(x)$, is both real and orthogonal. This completes the proof of Theorem 7.2.

7.4. Bilinear invariants

The classification which we described on p. 171 is closely related to the notion of a bilinear invariant.

Definition 7.4. *Let $A(x)$ be a representation of a group G. The invertible (complex) matrix L is said to be a bilinear invariant of $A(x)$ if*

$$A(x)LA'(x) = L. \tag{7.30}$$

Note the slight, but important, difference between (7.14) and (7.30). Every representation of a finite group has an Hermitian invariant; but a bilinear invariant does not always exist, as we shall see presently. The terminology is explained as follows: we associate with L the bilinear form

$$\beta(\mathbf{u}, \mathbf{v}) = \mathbf{u}L\mathbf{v}', \tag{7.31}$$

where $\mathbf{u}$ and $\mathbf{v}$ are variable row vectors. Then

$$\beta(\mathbf{u}A(x), \mathbf{v}A(x)) = \mathbf{u}A(x)LA'(x)\mathbf{v}'.$$

Therefore if (7.30) holds we have that

$$\beta(\mathbf{u}A(x), \mathbf{v}A(x)) = \beta(\mathbf{u}, \mathbf{v}).$$

It is in this sense that β, or its matrix L, can be called a bilinear invariant of $A(x)$.

Proposition 7.3. *A representation of a finite group possesses a bilinear invariant if and only if its character is real.*

Proof. If $\phi(x)$ is the character of

$$A(x) \quad (x \in G), \tag{7.32}$$

then the contragredient representation (p. 16)

$$A^\dagger(x) = (A(x^{-1}))' \quad (x \in G) \tag{7.33}$$

has character $\phi(x^{-1})$ which, by (2.37), is equal to $\bar{\phi}(x)$. Hence ϕ is real if and only if the representations (7.32) and (7.33) are equivalent (see p. 52), that is if there exists an invertible matrix L such that

$$L^{-1}A(x)L = (A(x^{-1}))' = (A(x))'^{-1}. \tag{7.34}$$

This equation can be written as

$$A(x)LA'(x) = L,$$

which means that L is a bilinear invariant of A.

More precise information about bilinear invariants can be obtained if we suppose that the representation is irreducible over $\mathbb{C}$.

Theorem 7.3. *Let F be a (complex) irreducible representation of a finite group G, with character χ.*

  (i) *F possesses a bilinear invariant if and only if χ is real;*
 (ii) *if L_1 and L_2 are bilinear invariants of F, then $L_2 = kL_1$, where k is a non-zero (complex) number;*
(iii) *if L is a bilinear invariant of F, then either L is symmetric ($L = L'$) or else skew-symmetric ($L = -L'$).*

Proof. (i) This is a special case of Proposition 7.3.

(ii) Suppose that L_1 and L_2 are bilinear invariants of F. Then by (7.34)

$$L_1^{-1}F(x)L_1 = L_2^{-1}F(x)L_2 = F^\dagger(x) \quad (x \in G).$$

Hence

$$F(x)(L_1 L_2^{-1}) = (L_1 L_2^{-1})F(x).$$

By the Corollary to Schur's Lemma (p. 27)

$$L_1 L_2^{-1} = kI,$$

where k is a scalar, which is clearly non-zero.

(iii) By hypothesis

$$F(x)LF'(x) = L.$$

Transposing this equation we have that

$$F(x)L'F'(x) = L'.$$

178

Thus if L is a bilinear invariant of F, so is L'. It follows from (ii) above that

$$L' = cL$$

and by transposition

$$L = cL'.$$

On eliminating L' between these equations we find that

$$L = c^2 L.$$

Therefore $c^2 = 1$, so that either $c = 1$ or $c = -1$.

We return to the classification mentioned on p. 171. In the next theorem we shall identify the type of an irreducible representation by examining its bilinear invariants, if any.

Theorem 7.4. *Let F be an irreducible (complex) representation of a finite group G.*

  (i) *F is of type I, if and only if it has a (complex) symmetric bilinear invariant;*

  (ii) *F is of type II, if and only if it has a (complex) skew-symmetric bilinear invariant;*

  (iii) *F is of type III, if and only if it has no bilinear invariant; more precisely: if*

$$F(x)LF'(x) = L \quad (x \in G).$$

  then $L = 0$.

Proof. (i) Suppose that F is of type I. Thus there exists an invertible matrix T such that

$$T^{-1}F(x)T = E(x) \quad (x \in G) \tag{7.35}$$

where $E(x)$ is real. Let

$$Q = \sum_{y \in G} E(y)E'(y).$$

Clearly Q is a real symmetric matrix which is positive definite and therefore invertible. As in (7.17) it can be shown that

$$E(x)QE'(x) = Q \quad (x \in G).$$

Substituting for $E(x)$ from (7.35) we obtain, after a short calculation, that

$$F(x)LF'(x) = L \quad (x \in G), \tag{7.36}$$

179

where

$$L = TQT'.$$

Evidently, L is an invertible (complex) symmetric matrix. Thus (7.36) establishes the fact that F has a symmetric bilinear invariant.

Conversely, suppose that (7.36) holds, where L is an invertible symmetric matrix. By Theorem 7.1, there exists an invertible matrix P such that

$$L = PP'.$$

We can therefore rewrite (7.36) as

$$(P^{-1}F(x)P)(P^{-1}F(x)P)' = I.$$

Thus the representation

$$R(x) = P^{-1}F(x)P,$$

which is equivalent to $F(x)$, is a (complex) orthogonal representation. By Theorem 7.2, $R(x)$ is equivalent to a real representation, which, in turn, is equivalent to $F(x)$. Thus $F(x)$ is of type I.

(ii) Suppose that $F(x)$ is of type II. Then

$$F(x) \sim \bar{F}(x);$$

so the character of $F(x)$ is real. By Theorem 7.3(iii), $F(x)$ has a bilinear invariant which must be either symmetric or skew-symmetric. But it cannot be symmetric, because this would imply that $F(x)$ is of type I, as we have just seen. Hence $F(x)$ has a skew-symmetric invariant.

Conversely, suppose that $F(x)$ has a skew-symmetric invariant. Hence it is not of type I, because it cannot also have a symmetric invariant (Theorem 7.3(ii)). Since F has a bilinear invariant, its character is real. Therefore

$$F \sim \bar{F}.$$

Thus F is of type II.

(iii) Let χ be the character of F. Then both $\bar{F}$ and $F^{\dagger}$ (see (7.33)) have the character $\bar{\chi}$ and are therefore equivalent. Now F is of type III if and only if $F \not\sim \bar{F}$, that is $F \not\sim F^{\dagger}$. Hence F and $F^{\dagger}$ are inequivalent irreducible representations. By Schur's Lemma the only solution of

$$F(x)L = LF^{\dagger}(x) \quad (x \in G)$$

is $L = 0$.

7.5. The character criterion

In the preceding section it was shown how the type of an irreducible representation can be determined by its bilinear invariant. However, this result is of little value from a practical point of view, since the definition of a bilinear invariant involves a large number of equations. It is therefore a remarkable achievement of Frobenius and Schur to have discovered a criterion which consists of a simple formula expressed solely in terms of the character.

Theorem 7.5. *Let F be a (complex) irreducible representation of a group G of order g, and let χ be the character of F.*
  Define

$$c = c(F) - 1 \quad \text{or} \quad -1 \quad \text{or} \quad 0, \tag{7.36}$$

according as F is of type I or II or III respectively. Then

$$c = (1/g) \sum_{y \in G} \chi(y^2). \tag{7.37}$$

Proof. Suppose that

$$F(x) = (f_{ij}(x))$$

is of degree f. We introduce an $f \times f$ matrix

$$U = (u_{\alpha\beta})$$

whose elements $u_{\alpha\beta}$ are independent variables, and we construct the matrix

$$V = \sum_{y \in G} F(y) U F'(y). \tag{7.38}$$

As before, we verify that

$$F(x) V F'(x) = V \quad (x \in G),$$

or

$$F(x) V = V F^{\dagger}(x).$$

Since F and $F^{\dagger}$ are irreducible representations, Schur's Lemma (p. 24) tells us that V is either the zero matrix, or else is invertible; in the latter case it is a bilinear invariant of F. Applying Theorem 7.4 we conclude that, if F is of type III, then $V = 0$; but when F is of type I or II, we have that $V' = V$ or $V' = -V$ respectively, whether V is zero or not. These results may be summarised by the statement that

$$V' = cV \tag{7.39}$$

for all choices of U, where c is as defined in (7.36). Substituting (7.38) in (7.39) we obtain that

$$\sum_{y \in G} F(y) U F'(y) = c \sum_{y \in G} F(y) U' F'(y),$$

whence on comparing the (i, j)th element on both sides

$$\sum_{y \in G} f_{i\alpha}(y) u_{\alpha\beta} f_{j\beta}(y) = c \sum_{y \in G} f_{i\alpha}(y) u_{\beta\alpha} f_{j\beta}(y).$$

Equating coefficients of $u_{\alpha\beta}$ on the left and right, we find that

$$\sum_{y \in G} f_{i\alpha}(y) f_{j\beta}(y) = \sum_{y \in G} f_{i\beta}(y) f_{j\alpha}(y),$$

where i, j, α and β each range from 1 to f. In particular, we may put $\beta = i$ and $\alpha = j$ and then sum over i and j. Thus

$$\sum_{i,j} \sum_{y \in G} f_{ij}(y) f_{ji}(y) = c \sum_{i,j} \sum_{y \in G} f_{ii}(y) f_{jj}(y). \tag{7.40}$$

Now

$$\sum_j f_{ij}(y) f_{ji}(y)$$

is the (i, j)th element of the matrix

$$(F(y))^2 = F(y^2).$$

Hence the left-hand side of (7.40) is equal to

$$\sum_{y \in G} \chi(y^2),$$

whilst the right-hand side becomes

$$\sum_{y \in G} \chi(y) \chi(y). \tag{7.41}$$

For every character we have that

$$\chi(y) = \bar{\chi}(y^{-1}).$$

Now (7.41) can be written as $g\langle \chi, \bar{\chi} \rangle$, and (7.40) reduces to

$$\sum_{y \in G} \chi(y^2) = cg\langle \chi, \bar{\chi} \rangle.$$

When F is of type III, the characters χ and $\bar{\chi}$ are distinct, whence $\langle \chi, \bar{\chi} \rangle = 0$. When F is of type I or II, we have that $\chi = \bar{\chi}$ and therefore $\langle \chi, \bar{\chi} \rangle = 1$. Thus

$$\sum_{y \in G} \chi(y^2) = cg$$

in all cases, as asserted.

Representations of type II are rather rare. A simple example is furnished by the two-dimensional irreducible representation of the quaternion group Q (see p. 62). For convenience we repeat here the relevant information:

$$Q = \mathrm{gp}\{a, b \mid a^4 = 1, a^2 = b^2, b^{-1}ab = a^3\}$$

	1	a^2	$\{a, a^2\}$	$\{b, a^2b\}$	$\{ab, a^3b\}$
χ	2	-2	0	0	0

Table 7.1.

In the group Q we have that

$$x^2 = \begin{cases} 1 & \text{if } x = 1 \quad \text{or} \quad a^2 \\ a^2 & \text{otherwise.} \end{cases}$$

Hence

$$\sum_{y \in G} \chi(y^2) = 2(2) + 6(-2) = -8,$$

which shows that $c = -1$.

As we observed on p. 62 the function given in Table 7.1 is also a character of the dihedral group

$$D_4 = \mathrm{gp}\{a, b \mid a^4 = b^2 = 1, b^{-1}ab = a^3\}.$$

But the corresponding representation of D_4 is of type I. For we now have that

$$x^2 = \begin{cases} a^2 & \text{if } x = a \quad \text{or} \quad a^3 \\ 1 & \text{otherwise.} \end{cases}$$

Hence

$$\sum_{y \in G} \chi(y^2) = 6(2) + 2(-2) = 8$$

so that $c = 1$ (see Exercise 12 of Chapter 2).

183

Exercises

1. The *Principal Axis Theorem* asserts that if A is a real $n \times n$ symmetric matrix, there exists a real orthogonal matrix R such that

$$R'AR = \text{diag}(\lambda_1, \lambda_2, \ldots, \lambda_n),$$

where $\lambda_1, \lambda_2, \ldots, \lambda_n$ are the latent roots of A. Deduce that, if A is positive definite, there exists a real matrix P such that

$$A = PP'.$$

2. Let $\varepsilon = \exp(\pi i/n)$. Show that the matrices

$$A = \begin{pmatrix} \varepsilon & 0 \\ 0 & \varepsilon^{-1} \end{pmatrix}, \quad B = \begin{pmatrix} 0 & 1 \\ -1 & 0 \end{pmatrix}$$

generate a representation of the *dicyclic* group

$$\Delta_n = \text{gp}\{a, b \,|\, a^{2n} = 1, \, a^n = (ab)^2 = b^2\}$$

by virtue of the correspondence $a \to A$, $b \to B$. Prove that this representation is irreducible over $\mathbb{C}$. By finding a skew-symmetric invariant, or otherwise, show that the representation is of type II.

3. Prove that an irreducible representation of type II is necessarily of even order.

4. Let F be an irreducible representation of degree f of a group G of order g, and let χ be the character of F. Prove that

$$\sum_{y \in G} F(y^2) = (cg/f)I,$$

where c is as defined in (7.36). Also show that

$$\sum_{y \in G} \chi(xy^2) = (cg/f)\chi(x) \quad (x \in G). \tag{7.42}$$

In the remaining exercises we suppose that

$$F^{(1)}, F^{(2)}, \ldots, F^{(k)}$$

is a complete set of inequivalent representations of a finite group G of order g. The character of $F^{(i)}$ is denoted by $\chi^{(i)}$, its degree by $f^{(i)}$, and we put $c_i = c(F^{(i)})$ in accordance with (7.36) $(i = 1, 2, \ldots, k)$.

5. For a fixed element u of G, let $\zeta(u)$ be the number of solutions of the equation

$$x^2 = u.$$

Prove that

$$\zeta(u) = \sum_{i=1}^{k} c_i \chi^{(i)}(u).$$

Deduce that the sum of the degrees of the representations of type I is greater than the sum of the degrees of the representations of type II.

6. By applying (7.42), or otherwise, prove that

$$\sum_{y_1 \ldots y_n} \chi(y_1^2 y_2^2 \ldots y_n^2) = c^n g^n f^{1-n} \quad (n \geqslant 1), \tag{7.43}$$

where the summation is extended over all sets of elements $y_1, y_2, \ldots, y_n$ ranging independently over G.

Let $\zeta_1(n)$ be the number of solutions of

$$y_1^2 y_2^2 \ldots y_n^2 = 1 \quad (n \geqslant 1).$$

Show that

$$\zeta_1(n) = g^{n-1} \sum_{i=1}^{k} c_i^n (f^{(i)})^{2-n}. \tag{7.44}$$

Deduce that the number of solutions of the equation

$$x^2 = y^2 \quad (x, y \in G)$$

is equal to the number of irreducible real characters.

APPENDIX

A.1. A generalisation of Vandermonde's determinant

The following simple result will help to close some of the gaps left in the main text.

Proposition A.1. Let $x_1, x_2, \ldots, x_n$ and t be indeterminates. Suppose that

$$g_i(t) = \sum_{k=1}^{n} a_{ik} t^{n-k} \quad (i = 1, 2, \ldots, n)$$

are n polynomials of degree less than n. Then

$$\det(g_i(x_j)) = \Delta(x) \det A, \tag{A.1}$$

where

$$A = (a_{ik}) \quad (i, k = 1, 2, \ldots, n). \tag{A.2}$$

Proof. We have that

$$g_i(x_j) = \sum_{k=1}^{n} a_{ik} x_j^{n-k}.$$

Hence by matrix multiplication

$$(g_i(x_j)) = A(x_j^{n-k}).$$

On taking determinants and using (4.34) we obtain (A.1).

As a first application consider the n polynomials

$$\phi_i(t) = t(t-1) \ldots (t-n+i+1), \quad (1 \leq i \leq n-1)$$

and

$$\phi_n(t) = 1.$$

Note that ϕ_i is a monic polynomial of degree $n - i$. In the present case the matrix A in (A.2) becomes

$$A = \begin{pmatrix} 1 & a_{12} & \ldots & a_{1n} \\ 0 & 1 & \ldots & a_{2n} \\ 0 & 0 & \ldots & 1 \end{pmatrix}, \tag{A.3}$$

186

so that $\det A = 1$. It follows that

$$\det(\phi_i(x_j)) = \Delta(x). \tag{A.4}$$

Now let

$$l_1 > l_2 > \ldots > l_n \geq 0$$

be a set of non-negative integers and put

$$e_i = n - i \quad (i = 1, 2, \ldots, n).$$

We then have that

$$\phi_i(l_j) = \frac{l_j!}{(l_j - e_i)!}.$$

Substituting in (A.4) and transposing the determinant we deduce that

$$\det\left(\frac{l_i!}{(l_i - e_j)!}\right) = \Delta(l),$$

which is the result required for Frobenius's degree formula (4.51).

Next, we turn our attention to the Cauchy determinant

$$C_0 = \det\left(\frac{1}{1 - x_i y_j}\right) \quad (i, j = 1, 2, \ldots, n).$$

Put

$$f(t) = \prod_{k=1}^{n} (1 - t x_k) = 1 - c_1 t + c_2 t - \ldots + (-1)^n c_n t^n,$$

where $c_1, c_2, \ldots, c_n$ are the elementary symmetric functions of $x_1, x_2, \ldots, x_n$ and let

$$f_i(t) = f(t)/(1 - t x_i) \quad (i = 1, 2, \ldots, n).$$

We regard the indeterminates x temporarily as parameters, which are incorporated in the coefficients. Thus we write

$$f_i(t) = \sum_{r=1}^{n} b_{ir}(x) t^{n-r},$$

which is a polynomial of degree $n - 1$ in t. The least common denominator of the jth column of C_0 is $f(y_j)$. On extracting all the denominators we obtain that

$$C_0 = \left(\prod_j f(y_j)\right)^{-1} \det(f_i(y_j)). \tag{A.5}$$

187

By Proposition A.1,

$$\det\big(f_i(y_j)\big) = \det\big(b_{ir}(x)\big)\Delta(y), \tag{A.6}$$

and it remains to evaluate $\det\big(b_{ir}(x)\big)$. Expanding $(1-tx_i)^{-1}$ as a geometric series we have that

$$f_i(t) = \left(\sum_{p=0}^{n} (-1)^p c_p t^p \right)\left(\sum_{q=0}^{\infty} x_i^q t^q \right),$$

whence comparing coefficients of t^{n-r} we find that

$$b_{ir}(x) = \sum_{p=0}^{n-r} (-1)^p c_p x_i^{n-r-p} = \psi_r(x_i),$$

say, where

$$\psi_r(t) = \sum_{p=0}^{n-r} (-1)^p c_p t^{n-r-p} \tag{A.7}$$

is a monic polynomial of degree $n-r$. Again, as in (A.3), the coefficient matrix A of (A.7) has the upper triangular form so that $\det A = 1$. Hence

$$\det\big(b_{ir}(x)\big) = \det\big(\psi_r(x_i)\big) = \Delta(x).$$

Finally, substituting (A.6) in (A.5) and noting that

$$\prod_j f(y_j) = \prod_{i,j} (1 - x_i y_j),$$

we find that

$$C_0 = \Delta(x)\Delta(y) \prod_{i,j} (1 - x_i y_j)^{-1},$$

as stated on p. 118.

A.2. The alternant quotient

Our aim in this section is to establish the formula (4.68), namely

$$\frac{\det(x_j^{n-i+p_i})}{\det(x_j^{n-r})} = \det(w_{j-i+p_i}), \tag{A.8}$$

where $p: p_1 \geqslant p_2 \geqslant \ldots \geqslant p_n \geqslant 0$ is a partition of n. The argument is based on a remark about recurrence relations, which is almost self-evident.

Lemma. *Let n be a positive integer. Suppose that the sequence*

$$z_0, z_1, z_2, \ldots$$

satisfies the recurrence relations

$$z_s = a_1 z_{s-1} + a_2 z_{s-2} + \ldots + a_n z_{s-n} \quad (s \geq n), \tag{A.9}$$

where

$$a_1, a_2, \ldots, a_n \tag{A.10}$$

are given. For fixed k and $r = 1, 2, \ldots, n$ define $\gamma(k, r)$ as follows:

$$\gamma(k, r) = \delta_{k, n-r} \quad \text{when } 0 \leq k \leq n - 1 \tag{A.11}$$

and, recursively,

$$\gamma(k, r) = \sum_{\lambda=1}^{n} a_\lambda \gamma(k - \lambda, r) \quad \text{when} \quad k \geq n. \tag{A.12}$$

Then

$$z_k = \sum_{r=1}^{n} \gamma(k, r) z_{n-r} \quad (k = 0, 1, 2, \ldots), \tag{A.13}$$

Thus the sequence (z_k) is completely determined by its first n members, the coefficients $\gamma(k, r)$ being well-defined polynomials in the set (A.10).

Proof. When $0 \leq k \leq n - 1$, equation (A.13) is a trivial consequence of (A.11).

Using induction on $k (\geq n)$ we deduce from (A.12) that

$$\sum_{r=1}^{n} \gamma(k, r) z_{n-r} = \sum_{\lambda=1}^{n} a_\lambda \sum_{r=1}^{n} \gamma(k - \lambda, r) z_{n-r} = \sum_{\lambda} a_\lambda z_{k-\lambda} = z_k,$$

by virtue of (A.9). Hence (A.13) holds in all cases.

We are going to apply this lemma to several sequences with a common recurrence relation. Consequently, all these sequences will satisfy (A.13) with the same coefficients $\gamma(k, r)$.

Let $x_1, x_2, \ldots, x_n$ and t be indeterminates. Then

$$\prod_{j=1}^{n} (t - x_j) = t^n - c_1 t^{n-1} + c_2 t^{n-2} - \ldots + (-1)^n c_n,$$

where

$$c_1, c_2, \ldots, c_n$$

are the elementary symmetric functions of $x_1, x_2, \ldots, x_n$. For a fixed value of j, put $t = x_j$. It follows that

$$x_j^n = c_1 x_j^{n-1} - c_2 x_j^{n-2} + \ldots + (-1)^{n-1} c_n. \tag{A.14}$$

More generally, when $s \geq n$, we multiply (A.14) by x_j^{s-n} and obtain that

$$x_j^s = \sum_{r=1}^{n} (-1)^{r-1} c_r x_j^{s-r}. \tag{A.15}$$

Hence the sequence $z_k = x_j^k$ ($k = 0, 1, \ldots$) satisfies (A.9) with

$$a_r = (-1)^{r-1} c_r \quad (r = 1, 2, \ldots, n). \tag{A.16}$$

Choosing for k in turn the integers

$$n - i + p_i \quad (i = 1, 2, \ldots, n)$$

we deduce from (A.13) that

$$x_j^{n-i+p_i} = \sum_{r=1}^{n} \gamma(n - i + p_i, r) x_j^{n-r},$$

where the coefficients $\gamma(k, r)$ have been constructed with the aid of (A.16). This equation may conveniently be written in matrix form, thus

$$(x_j^{n-i+p_i}) = (\gamma(n - i + p_i, r)) (x_j^{n-r}).$$

Taking determinants we find that

$$\frac{\det(x_j^{n-i+p_i})}{\det(x_j^{n-r})} = \det(\gamma(n - i + p_i, r)). \tag{A.17}$$

We observe that the right-hand side depends only on $c_1, c_2, \ldots, c_n$ in accordance with the Fundamental Theorem on symmetric functions.

Next, we consider the sequence

$$z_k = w_{j-n+k} \quad (k = 0, 1, \ldots), \tag{A.18}$$

where j is a fixed integer such that $1 \leq j \leq n$. When $k = s \geq n$, the suffix of w is positive and we deduce from (4.63) that

$$w_{j-n+s} = \sum_{r=1}^{n} (-1)^{r-1} c_r w_{j-n+s-r}.$$

This is the same recurrence relation as (A.15). Substituting (A.18) in (A.13) we find that

$$w_{j-n+k} = \sum_{r=1}^{n} \gamma(k, r) w_{j-r}.$$

On putting $k = n - i + p_i$ ($i = 1, 2, \ldots, n$) we obtain the matrix equation

$$(w_{j-i+p_i}) = (\gamma(n - i + p_i, r))(w_{j-r}).$$

Now (w_{j-r}) is an upper triangular matrix whose diagonal elements are equal to $w_0 (=1)$. Hence

$$\det(w_{j-r}) = 1,$$

and it follows that

$$\det(w_{j-i+p_i}) = \det(\gamma(n-i+p_i, r)).$$

Comparison with (A.17) completes the proof of (A.8).

A.3. Jacobi's Theorem on inverse matrices

The notation for submatrices has already been explained on p. 133. The determinant of a submatrix is called a minor. Jacobi's Theorem deals with minors drawn from a pair of inverse matrices.

Theorem A.1. *Let W and A be a pair of $m \times m$ inverse matrices, that is*

$$WA = I. \tag{A.19}$$

Suppose that $W(\lambda; \mu)$ is the $n \times n$ submatrix of W formed by the intersections of the rows labelled by

$$\lambda: 1 \leqslant \lambda_1 < \lambda_2 < \ldots \lambda_n \leqslant m$$

and the columns labelled by

$$\mu: 1 \leqslant \mu_1 < \mu_2 < \ldots < \mu_n \leqslant m.$$

Assuming that $n < m$ denote by

$$\rho: 1 \leqslant \rho_1 < \rho_2 < \ldots < \rho_{m-n} \leqslant m$$

and

$$\sigma: 1 \leqslant \sigma_1 < \sigma_2 < \ldots < \sigma_{m-n} \leqslant m$$

the index sets that are complementary to λ and μ respectively in the set

$$1, 2, \ldots, m.$$

Then

$$\det W(\lambda; \mu) \det A = (-1)^\varepsilon \det A(\sigma; \rho), \tag{A.20}$$

where

$$\varepsilon = \sum_i (\lambda_i + \mu_i). \tag{A.21}$$

191

Proof. (i) We consider the special case, in which the index set

$$\nu : 1 < 2 < \ldots < n$$

and its complement

$$\omega : n+1 < n+2 < \ldots < m$$

are involved. Partition W and A as

$$\begin{pmatrix} W_1 & W_3 \\ W_4 & W_2 \end{pmatrix} \quad \text{and} \quad \begin{pmatrix} A_1 & A_3 \\ A_4 & A_2 \end{pmatrix}, \tag{A.22}$$

where W_1 and A_1 are the leading square matrices of degree n, while W_2 and A_2 are square matrices of degree $m-n$. Let I_1 and I_2 be the unit matrices of degrees n and $m-n$ respectively. By substituting (A.22) in (A.19) we readily find that

$$\begin{vmatrix} W_1 & W_3 \\ 0 & I_2 \end{vmatrix} \begin{vmatrix} A_1 & A_3 \\ A_4 & A_2 \end{vmatrix} = \begin{vmatrix} I_1 & 0 \\ A_4 & A_2 \end{vmatrix}.$$

For example, we have that $W_1 A_3 + W_3 A_2 = 0$. On taking determinants we deduce that

$$\det W_1 \det A = \det A_2, \tag{A.23}$$

which in the general notation for submatrices becomes

$$\det W(\nu; \nu) \det A = \det A(\omega; \omega).$$

This is indeed an instance of Jacobi's Theorem, because in the present case (A.21) reduces to

$$\varepsilon = 2 \sum_i \nu_i,$$

which is an even integer.

(ii) In order to pass to the general case we permute the rows and columns of W and A in such a way that the submatrix selected by λ and μ occupies the leading $n \times n$ position. Thus we introduce the permutations

$$\phi = \begin{pmatrix} 1 & 2 & \ldots & n & n+1 & \ldots & m \\ \lambda_1 & \lambda_2 & \ldots & \lambda_n & \rho_1 & \ldots & \rho_{m-n} \end{pmatrix} \tag{A.24}$$

and

$$\psi = \begin{pmatrix} 1 & 2 & \ldots & n & n+1 & \ldots & m \\ \mu_1 & \mu_2 & \ldots & \mu_n & \sigma_1 & \ldots & \sigma_{m-n} \end{pmatrix}.$$

Let P and Q be the permutation matrices that correspond to ϕ and ψ, that is

$$P = (\delta_{i\phi,j}), \qquad Q = (\delta_{i\psi,j}).$$

If we wish to permute the rows of W in accordance with ϕ, we have to replace W by PW. On the other hand, in order to permute the columns of W in accordance with ψ, we must change W into WQ'. For, acting on the columns of W is equivalent to acting on the rows of W', after which columns and rows must again be interchanged. Thus W has to be replaced by $(QW')' = WQ'$. If the permutations ϕ and ψ operate simultaneously on the rows and columns respectively, we obtain the matrices

$$\tilde{W} = PWQ' \quad \text{and} \quad \tilde{A} = PAQ'.$$

Using (A.19) and the fact that $Q'Q = PP' = I$, we find that

$$\tilde{W}\tilde{A}' = I.$$

Hence we may apply (A.23) to the pair $(\tilde{W}, \tilde{A}')$. Partitioning these matrices analogously to (A.22), we write, with a self-explanatory notation,

$$\tilde{W} = \begin{pmatrix} (w_{\lambda\mu}) & (w_{\lambda\sigma}) \\ (w_{\rho\mu}) & (w_{\rho\sigma}) \end{pmatrix},$$

$$\tilde{A} = \begin{pmatrix} (a_{\mu\lambda}) & (a_{\mu\rho}) \\ (a_{\sigma\lambda}) & (a_{\sigma\rho}) \end{pmatrix},$$

and we deduce that

$$\det W(\lambda, \mu) \det \tilde{A}' = \det A(\sigma, \rho). \tag{A.25}$$

Now

$$\det \tilde{A}' = \det(QA'P') = \det A \, \det P \, \det Q,$$

and it remains to evaluate $\det P \det Q$.

(iii) Generally, if P is the permutation matrix that corresponds to ϕ, then

$$\det P = \zeta(\phi),$$

where ζ is the alternating character (p. 4). For $\det P$ is clearly a linear character of the symmetric group, and as it is not identically equal to unity, it must coincide with ζ. We determine $\zeta(\phi)$ by counting the number of inversions in (A.24), [**13**, p. 134]. If $\lambda_1 = 1$, no inversions are caused by λ_1; but if $\lambda_1 > 1$, then $\lambda_1 - 1$ inversions are caused, since λ_1 precedes the smaller integers $1, 2, \ldots, \lambda_1 - 1$. Hence in all cases

$$\lambda_1 - 1 \quad \textit{inversions are due to } \lambda_1.$$

Next, if $\lambda_2 = 2$, then necessarily $\lambda_1 = 1$, and no inversions are caused by λ_2; but if $\lambda_2 > 2$, then $\lambda_2 - 2$ inversions are caused, since λ_2 precedes the smaller integers $1, 2, \ldots, \lambda_2 - 1$, with the exception of λ_1. Hence in all cases

$$\lambda_2 - 2 \quad \textit{inversions are due to } \lambda_2.$$

Continuing in this manner we conclude that

$$\lambda_i - i \quad \textit{inversions are due to } \lambda_i$$

$(i = 1, 2, \ldots, n)$. No inversions arise in relation to $\rho_1, \rho_2, \ldots, \rho_{m-n}$, as these integers are arranged in increasing order of magnitude. It follows that the total number of inversions in ϕ is equal to

$$\sum_{i=1}^{n} (\lambda_i - i).$$

Similarly, the total number of inversions in ψ is given by

$$\sum_{i=1}^{n} (\mu_i - i).$$

Hence

$$\det P \det Q = \zeta(\phi)\, \zeta(\psi) = (-1)^{\varepsilon},$$

where

$$\varepsilon = \sum_i (\lambda_i + \mu_i - 2i) \equiv \sum_i (\lambda_i + \mu_i) \quad (\text{mod } 2).$$

Therefore,

$$\det \tilde{A}' = (-1)^{\varepsilon} \det A.$$

Substituting in (A.25) and transposing the factor $(-1)^{\varepsilon}$ we obtain (A.20), as required.

A.4. Quadratic forms

In this section we shall prove Theorem 7.1, that is we shall establish the 'sum of squares' reduction

$$q = \mathbf{x} A \mathbf{x}' = z_1^2 + z_2^2 + \ldots + z_r^2, \tag{A.26}$$

where

$$A = (a_{ij})$$

is a non-zero (complex) symmetric matrix, and

$$\mathbf{z} = \mathbf{x}P, \tag{A.27}$$

P being an invertible matrix. In general, the transformation (A.27) is the resultant of several, suitably chosen, linear transformations

$$x \to u \to v \to \ldots \to y \to z.$$

In particular, since a permutation of the variables may be regarded as a linear transformation, it is permissible to renumber the variables at any stage of the proof.

We use induction with respect to n. When $n = 1$, the quadratic form reduces to

$$q = a_{11}x_1^2 = z_1^2,$$

where

$$z = (\sqrt{a_{11}})x_1$$

is an invertible transformation because $a_{11} \neq 0$. We shall now suppose that $n \geqslant 2$ and that the theorem holds for not more than $n - 1$ variables. Two cases have to be distinguished.

(i) Suppose that not all diagonal entries of A are zero. After permuting the variables, if necessary, we may assume that

$$a_{11} \neq 0.$$

We define a quadratic form q_1 by the equation

$$q_1 = q - \frac{1}{a_{11}}(a_{11}x_1 + a_{12}x_2 + \ldots + a_{1n}x_n)^2. \tag{A.28}$$

It is easy to verify that the variable x_1 does not occur in q_1, the terms $a_{11}x_1^2$ and $2a_{1k}x_1x_k$ $(k > 1)$ having been cancelled on the right. If q_1 happens to be the zero form, then

$$q = z_1^2, \tag{A.29}$$

where

$$z_1 = \frac{1}{\sqrt{a_{11}}}(a_{11}x_1 + a_{12}x_2 + \ldots + a_{1n}x_n)$$

$$z_k = x_k \quad (k \geqslant 2) \tag{A.30}$$

is an invertible transformation from $\mathbf{x}$ to $\mathbf{z}$; thus the reduction is given by (A.29). But if q_1 is not zero, the inductive hypothesis states that there is an invertible transformation

$$(x_2, x_3, \ldots, x_n) \rightarrow (z_2, z_3, \ldots, z_n) \qquad \text{(A.31)}$$

such that

$$q_1 = z_2^2 + z_3^2 + \ldots + z_r^2,$$

where the number of squares does not exceed $n-1$. On defining z_1 as in (A.30) we augment (A.31) to a transformation

$$\mathbf{x} \rightarrow \mathbf{z},$$

with the desired property (A.26).

(ii) Suppose next that

$$a_{11} = a_{22} = \ldots = a_{nn} = 0.$$

Since $A \neq 0$, there exists at least one non-zero entry in it, say a_{ij}, where $1 \leqslant i < j \leqslant n$. After permuting the variables appropriately we may assume that

$$a_{12} \neq 0.$$

We now make the preliminary transformation

$$y_1 = x_1$$

$$y_2 = -x_1 + x_2$$

$$y_k = x_k \quad (k = 3, 4, \ldots, n),$$

the last equation being omitted when $n=2$. When expressed in terms of $\mathbf{y}$, the form q is given by

$$q = 2a_{12}y_1(y_1 + y_2) + \ldots = 2a_{12}y_1^2 + \ldots.$$

Thus the coefficient of y_1^2 is non-zero, and we can proceed with the reduction as in (i) above. This complexes the proof of the theorem.

As was remarked on p. 172, the 'sum of pure squares' result holds also for positive definite quadratic forms over the real field. For in that case $a_{kk} > 0$ ($k = 1, 2, \ldots, n$) so that the transformation (A.30) has real coefficients. Moreover, if q is positive definite so is the form q_1 in (A.28). For, if not, we could find real numbers

$$u_2, u_3, \ldots, u_n,$$

not all zero, such that

$$q_1(u_2, u_3, \ldots, u_n) \leqslant 0.$$

196

Defining u_1 by

$$u_1 = -\frac{1}{a_{11}}(a_{12}u_2 + \ldots + a_{1n}u_n)$$

we should get that

$$q(u_1, u_2, \ldots, u_n) \leqslant 0$$

contradicting the hypotheses that q is positive definite. Thus we have established Proposition 7.1.

A.5. Congruence relations in an algebraic number field

Proposition 6.3 (p. 165) can be significantly generalised if we apply some simple results about algebraic number fields. This provides a deeper insight into the arithmetic properties of characters and leads to further useful devices for constructing character tables. We begin with a brief account of the necessary preliminaries. The symbol

$$o(x)$$

denotes the order (period) of an element in a group.

Proposition A.2. *Let x be an element of order n in a group. Suppose that*

$$n = p^\alpha m,$$

where p is a prime and $(p, m) = 1$. Then x can be uniquely factorised in the form

$$x = uv, \tag{A.32}$$

where

$$o(u) = p^\alpha, \quad o(v) = m, \quad uv = vu. \tag{A.33}$$

Proof. Since $(m, p) = 1$, there exist integers h and k such that

$$1 = hm + kp^\alpha. \tag{A.34}$$

Hence

$$x = (x^{hm})(x^{kp^\alpha}) = uv,$$

say, where

$$u = x^{hm}, \quad v = x^{kp^\alpha}. \tag{A.35}$$

It is obvious that u and v commute because they are powers of the same element x. Next, we have that

$$u^{p^\alpha} = 1.$$

197

Hence $o(u) = p^\beta$, *where* $\beta \leqslant \alpha$, and (A.35) implies that

$$u^{p^\beta} = x^{hmp^\beta} = 1.$$

Therefore $o(x)$ is a factor of hmp^β, that is

$$p^\alpha m \,|\, hmp^\beta, \quad p^\alpha \,|\, hp^\beta.$$

It follows from (A.34) that $(h, p) = 1$. Hence $p^\alpha \,|\, p^\beta$ and therefore $\alpha \leqslant \beta$. Thus, finally, $\alpha = \beta$, that is

$$o(u) = p^\alpha.$$

Similarly, it is proved that

$$o(v) = m.$$

In order to show that the decomposition (A.32) is unique, suppose that

$$x = u_1 v_1,$$

where u_1 and v_1 satisfy the conditions laid down in (A.35) for u and v. Hence

$$x^{mh} = u_1^{mh} v_1^{mh} = u_1^{mh} = u_1,$$

because $mh \equiv 1 \pmod{p^\alpha}$ by virtue of (A.34). Similarly

$$x^{kp^\alpha} = v_1.$$

Thus u_1 and v_1 are identical with u and v respectively. It is convenient to use the following

Definition A.1. *Let* x *be an element of a group and suppose that*

$$x = uv,$$

where u *and* v *satisfy the conditions* (A.33). *Then* u *and* v *are called the p-component and the p-regular component (p'-component) of* x *respectively.*

In § 5.1 the reader has already met some elementary facts about algebraic numbers. We recall that a number ξ is said to be algebraic if it is the root of a polynomial with coefficients in $\mathbb{Q}$, the field of rational numbers. This polynomial is unique if we postulate that its highest coefficient is equal to unity (monic polynomial) and that its degree is minimal; it is then called the minimal polynomial for ξ (over $\mathbb{Q}$). We shall denote it by

$$X(t) \quad (X(\xi) = 0). \tag{A.36}$$

The extension field $\mathbb{Q}(\xi)$ consists of all rational functions in ξ with rational coefficients [see **23**, Chapter 3]. If the minimal equation for ξ is of degree r, we write

$$r = [\mathbb{Q}(\xi) : \mathbb{Q}]$$

and we say that $\mathbb{Q}(\xi)$ is of degree r over $\mathbb{Q}$. Alternatively, $\mathbb{Q}(\xi)$ may be viewed as a vector space over $\mathbb{Q}$, and it turns out that its dimension is equal to r.

Now let η be another algebraic number and suppose that its minimal polynomial (over $\mathbb{Q}$)

$$Y(t) \quad (Y(\eta) = 0) \tag{A.37}$$

is of degree s, so that

$$s = [\mathbb{Q}(\eta) : \mathbb{Q}].$$

So far, we have regarded $\mathbb{Q}$ as the ground field to which all algebraic numbers are referred. However, it is possible to choose a different (larger) field $\mathbf{K}$ as ground field. In particular, we shall consider the case in which

$$\mathbf{K} = \mathbb{Q}(\eta).$$

Of course, ξ is algebraic over $\mathbf{K}$; indeed $X(t)$ may be regarded as having coefficients in $\mathbf{K}$ because $\mathbb{Q} \subset \mathbf{K}$. But, in general X will not be the minimal polynomial for ξ over $\mathbf{K}$. In fact, let

$$\Xi(t; \eta) \quad (\Xi(\xi; \eta) = 0) \tag{A.38}$$

be the minimal polynomial for ξ over $\mathbf{K}$, where the notation indicates that the coefficients of Ξ depend on η and suppose that Ξ is of degree σ. Then $\mathbf{K}(\xi)$ is an extension of $\mathbf{K}$ and we have that

$$\sigma = [\mathbf{K}(\xi) : \mathbf{K}].$$

Alternatively, we can describe $\mathbf{K}(\xi)$ as the field obtained from $\mathbb{Q}$ by adjoining both ξ and η, and we may write

$$\mathbb{Q}(\xi, \eta) = \mathbf{K}(\xi).$$

An important theorem on multiple extensions, which we have to quote here without proof, states that

$$[\mathbb{Q}(\xi, \eta) : \mathbb{Q}(\eta)][\mathbb{Q}(\eta) : \mathbb{Q}] = [\mathbb{Q}(\xi, \eta) : \mathbb{Q}], \tag{A.39}$$

where the right-hand side is the dimension of $\mathbb{Q}(\xi, \eta)$, regarded as a vector space over $\mathbb{Q}$.

We are now ready to enunciate the main result:

Theorem A.2 [**22a**, p. 76]. *Let F be a (complex) representation, which need*

199

not be irreducible, of a finite group G. Suppose that F is of degree f and has character χ.

Let x be an element of order n in G, where

$$n = p^\alpha m,$$

p being a prime and $(m, p) = 1$. *Write*

$$x = uv,$$

where u and v are the p-component and the p-regular component of x respectively.

Let

$$\mathbf{K} = \mathbb{Q}(\exp(2\pi i/m))$$

be the extension of $\mathbb{Q}$ *by a primitive mth root of unity. Assume that* $\chi(x)$ *happens to lie in* $\mathbf{K}$. *Then we have that*

$$\chi(x) \equiv \chi(v) \pmod{p}, \tag{A.40}$$

that is

$$\chi(x) - \chi(v) = p\omega,$$

where ω *is an algebraic integer in* $\mathbf{K}$.

Proof. As in the proof of Proposition A.2 we choose integers h and k such that (A.34) and (A.35) hold. The complex number

$$\varepsilon = \exp(2\pi i/n) \tag{A.41}$$

generates a cyclic group of order n. Hence ε can be split into a p-component and a p-regular component, say

$$\varepsilon = \xi\eta, \tag{A.42}$$

where

$$\xi = \varepsilon^{hm}, \quad o(\xi) = p^\alpha; \tag{A.43}$$

$$\eta = \varepsilon^{kp^\alpha}, \quad o(\eta) = m. \tag{A.44}$$

The field $\mathbf{K}$ was defined as $\mathbb{Q}(\eta_0)$ where

$$\eta_0 = \exp(2\pi i/m).$$

But we may equally well write

$$\mathbf{K} = \mathbb{Q}(\eta). \tag{A.45}$$

For

$$\eta = \varepsilon^{kp^\alpha} = \exp(2\pi i k p^\alpha/n) = \exp(2\pi i k/m),$$

that is

$$\eta = \eta_0^k, \tag{A.46}$$

whence

$$\mathbb{Q}(\eta) \subset \mathbb{Q}(\eta_0).$$

Conversely, by (A.44),

$$\eta^{p^\alpha} = \eta_0^{kp^\alpha} = \eta_0,$$

because

$$kp^\alpha \equiv 1 \pmod{m}.$$

Hence

$$\mathbb{Q}(\eta_0) \subset \mathbb{Q}(\eta),$$

which proves (A.45).

Next, we assert that

$$\mathbb{Q}(\xi, \eta) = \mathbb{Q}(\varepsilon). \tag{A.47}$$

For it is clear from (A.42) that

$$\mathbb{Q}(\varepsilon) \subset \mathbb{Q}(\xi, \eta),$$

while (A.43) and (A.44) imply that

$$\mathbb{Q}(\xi, \eta) \subset \mathbb{Q}(\varepsilon).$$

Substituting these results in (A.39) we have that

$$[\mathbb{Q}(\xi, \eta) : \mathbb{Q}(\eta)][\mathbb{Q}(\eta) : \mathbb{Q}] = [\mathbb{Q}(\varepsilon) : \mathbb{Q}]. \tag{A.48}$$

By Theorem 6.2,

$$[\mathbb{Q}(\varepsilon) : \mathbb{Q}] = \phi(n), \quad [\mathbb{Q}(\eta) : \mathbb{Q}] = \phi(m).$$

The multiplicative property of the Euler function [11, § 5.5] implies that

$$\phi(n) = \phi(p^\alpha m) = \phi(p^\alpha)\phi(m).$$

Hence we deduce from (A.48) that

$$[\mathbb{Q}(\xi, \eta) : \mathbb{Q}(\eta)] = \phi(p^\alpha).$$

Accordingly, if

$$\Xi(t; \eta) \quad (\Xi(\xi; \eta) = 0)$$

is the minimal polynomial for ξ over $\mathbb{Q}(\eta)$, then

$$\text{degree } \Xi = \phi(p^\alpha).$$

On the other hand, since ξ is a primitive (p^α)th root of unity, Theorem 6.2

tells us that ξ is of degree $\phi(p^\alpha)$ over $\mathbb{Q}$. Indeed its minimal polynomial over $\mathbb{Q}$ is the cyclotomic polynomial

$$\Phi(t) = \Phi_{p\alpha}(t) \quad (\Phi(\xi) = 0),$$

which was explicitly given in (6.9); in particular,

$$\text{degree } \Phi = \phi(p^\alpha). \tag{A.49}$$

Now Φ may also be regarded as a polynomial over $\mathbb{Q}(\eta)$ having ξ as a root. Hence Φ is divisible by the minimal polynomial Ξ. Since, in addition, Φ and Ξ are monic and of the same degree, they must be equal, thus

$$\Xi(t; \eta) = \Phi(t). \tag{A.50}$$

It is convenient to record the consequence of this statement in the following

Lemma. *Let* $\Theta(t; \eta)$ *be a monic polynomial over* $\mathbb{Q}(\eta)$ *such that*

$$\Theta(\xi; \eta) = 0. \tag{A.51}$$

Then there exists a polynomial $\Gamma(t; \eta)$ *such that*

$$\Theta(t; \eta) = \Phi(t)\Gamma(t; \eta). \tag{A.52}$$

Moreover, if the coefficients of Θ *are algebraic integers in* $\mathbb{Q}(\eta)$, *so are the coefficients of* Γ.

Proof. On dividing Θ by Φ we obtain an equation of the form

$$\Theta(t; \eta) = \Phi(t)\Gamma(t; \eta) + P(t; \eta),$$

where

$$\deg P < \deg \Phi.$$

By (A.49) and (A.51)

$$P(\xi; \eta) = 0.$$

But since Φ is the minimal polynomial for ξ, it follows that $P = 0$; this proves (A.52). Since Φ is monic, no denominators are introduced when Θ is divided by Φ. Hence if the coefficients of Θ are algebraic integers, so are those of Γ.

Continuing the proof of Theorem A.2 we observe that the latent roots of $F(x)$ are numbers of the form

$$\varepsilon^{a_1}, \varepsilon^{a_2}, \ldots, \varepsilon^{a_f},$$

where ε is given in (A.41) and $a_1, a_2, \ldots, a_f$ are rational integers. For brevity we shall write

$$F(x) \sim \mathrm{diag}(\varepsilon^{a_j}) \quad (1 \leqslant j \leqslant f),$$

whence by (A.42),

$$F(x) \sim \mathrm{diag}(\xi^{a_j} \eta^{a_j}).$$

We introduce the polynomial

$$\Lambda(t; \eta) = \sum_{j=1}^{f} \eta^{a_j} t^{a_j}. \tag{A.53}$$

Clearly

$$\chi(x) = \Lambda(\xi; \eta). \tag{A.54}$$

Let v be the p-regular component of x. Then by (A.35)

$$F(v) \sim \mathrm{diag}((\xi\eta)^{a_j k p^\alpha}) = \mathrm{diag}(\eta^{a_j k p^\alpha})$$

because $o(\xi) = p^\alpha$. But, by (A.34),

$$kp^\alpha \equiv 1 \pmod{m}$$

and, since $o(\eta) = m$, we obtain that

$$F(v) \sim \mathrm{diag}(\eta^{a_j}).$$

Hence

$$\chi(v) = \Lambda(1; \eta). \tag{A.55}$$

By hypothesis, $\chi(x)$ lies in $\mathbb{Q}(\eta)$. Hence the polynomial

$$\Theta(t; \eta) = \Lambda(t; \eta) - \chi(x)$$

has the property that

$$\Theta(\xi; \eta) = 0.$$

Moreover, the coefficients of Θ are algebraic integers. Hence, by the preceding lemma,

$$\Theta(t; \eta) = \Phi(t)\Gamma(t; \eta).$$

On putting $t = 1$ we get that

$$\chi(v) - \chi(x) = \Phi(1)\Gamma(1; \eta).$$

Inspection of (6.9) shows that

$$\Phi(1) = p,$$

while $\Gamma(1; \eta)$ is some algebraic integer ω, say. This completes the proof of the theorem.

The reader may find it illuminating to test the strength of the theorem by applying it to the character table of a symmetric group, where all entries are known to be rational integers.

Example. In the table S_5 (p. 137) the column headed (23) is the conjugacy class of

$$x = (12)(345),$$

which is an element of order 6. Since there are 20 elements in this class, we have that

$$\sum_{\chi} (\chi(x))^2 = \tfrac{120}{20} = 6.$$

Hence

$$|\chi(x)| \leqslant 2 \tag{A.56}$$

for all characters. We may take $p = 2$ or $p = 3$. The 2-regular component of x is (345), and its 3-regular component is (12). Hence the entries in the column headed (23) are congruent mod 3 to the corresponding entries in the column $(1^3 2)$ and congruent mod 2 to the entries in the column $(1^2 3)$. This information suffices to determine the column (23) mod 6; and finally (A.56) gives the precise values. For example

$$\chi^{[41]}(x) \equiv \begin{cases} 2 & \mod 3, \\ 1 & \mod 1, \end{cases}$$

whence

$$\chi^{[41]}(x) = -1.$$

It is an instructive exercise to continue Example 5 of Chapter 6 and to construct the complete character table of S_5 by means of Theorem A.2.

Description of group	Notation	Page
Cyclic of order 3	Z_3	53
Cyclic of order n	Z_n	53
Four-group	V	56
Finite Abelian		56
Symmetric of degree 3	S_3	50
Symmetric of degree 4	S_4	106
Symmetric of degree 5	S_5	137
Symmetric of degree 6	S_6	140
Alternating of degree 4	A_4	61
Alternating of degree 5	A_5	79
Dihedral of order 8	D_4	62
Quaternion of order 8	Q	62
Dihedral of order $4m$	D_{2m}	65–6
Dihedral of order $4m+2$	D_{2m+1}	66
Non-Abelian of order p^3 and exponent p^2		98
Non-Abelian of order p^3 and exponent p		98
Dicyclic of order 12		65
Non-Abelian of order 21		97

SOLUTIONS

Throughout the solutions, Exercise x of Chapter y is referred to as Exercise $y.x$.

Solutions to Exercises on Chapter 1

1. The element u of G belongs to the kernel of σ if and only if $Ht_iu = Ht_i$, that is $u \in t_i^{-1}Ht_i$ for all i ($i = 1, 2, \ldots, n$).

2. It suffices to verify that $\{C(a)\}^3 = I$. If the representation were reducible over the real field, there would exist a real matrix T such that

$$T^{-1}C(a)T = \begin{pmatrix} \alpha & 0 \\ \gamma & \beta \end{pmatrix},$$

where α, β and γ are real. Then α and β would be the latent roots of $C(a)$. However, the latent roots satisfy $\lambda^2 + \lambda + 1 = 0$ and are complex.

3. If $\mathbf{u}^{(1)}$, $\mathbf{u}^{(2)} \in U$ and a_1, a_2 are scalars, then $(a_1\mathbf{u}^{(1)} + a_2\mathbf{u}^{(2)})\varepsilon = a_1\mathbf{u}^{(1)} + a_2\mathbf{u}^{(2)}$. Hence U is a subspace. Also if $\mathbf{u} \in U$, then $\mathbf{u}\varepsilon \in U$ because $(\mathbf{u}\varepsilon)\varepsilon = \mathbf{u}\varepsilon^2 = \mathbf{u}\varepsilon = \mathbf{u}$. Therefore U is invariant under ε. Similarly for W, which is the kernel of ε.

For an arbitrary $\mathbf{v} \in V$ write

$$\mathbf{v} = \mathbf{v}\varepsilon + \mathbf{v}(\iota - \varepsilon).$$

It is easy to verify that $\mathbf{v}\varepsilon \in U$ and $\mathbf{v}(\iota - \varepsilon) \in W$. Hence $V = U + W$. But $U \cap W = \varnothing$, because $\mathbf{u} = \mathbf{w}$ implies that $\mathbf{u}\varepsilon = \mathbf{w}\varepsilon$, that is $\mathbf{u} = \mathbf{0}$.

Let $V = [\mathbf{v}_1, \mathbf{v}_2, \ldots, \mathbf{v}_m]$ be the space of m-dimensional row vectors. Define ε by $\mathbf{v}_i\varepsilon = \mathbf{v}_iE$ ($i = 1, 2, \ldots, m$). Choose a basis of V adapted to U and W, say

$$V = [\mathbf{u}_1, \ldots, \mathbf{u}_r, \mathbf{w}_1, \ldots, \mathbf{w}_{m-r}].$$

Relative to this basis ε is given by the matrix

$$J = \begin{pmatrix} I_r & 0 \\ 0 & 0 \end{pmatrix}.$$

If T is the matrix which describes the change of basis, then $T^{-1}ET = J$.

4. Let $E = A(1)$. Then $E^2 = E$. By Exercise 1.3 we have that $E \sim J$, and it may be assumed from the outset that $A(1) = J$. The result follows by writing

$$A(x) = \begin{pmatrix} B(x) & B_1(x) \\ B_2(x) & B_3(x) \end{pmatrix},$$

where $B(x)$ is an $r \times r$ matrix, and using the fact that

$$A(x) = A(1)A(x)A(1) \quad (x \in G).$$

5. If $\sigma \in G$, then $\mathbf{u}_j\sigma = \mathbf{v}_{j\sigma} - \mathbf{v}_{m\sigma} = (\mathbf{v}_{j\sigma} - \mathbf{v}_m) - (\mathbf{v}_{m\sigma} - \mathbf{v}_m)$, with the obvious interpretation when $j\sigma = m$ or $m\sigma = m$. Hence $\mathbf{u}_j\sigma \in U$, that is U is a G-module. The vectors $\mathbf{u}_1, \mathbf{u}_2, \ldots, \mathbf{u}_{m-1}$ are obviously linearly independent so that dim $U = m - 1$.

If A is the matrix representation afforded by U, with character ϕ we find that

$$A(\tau) = \begin{pmatrix} 0 & 1 & 0 \\ 1 & 0 & 0 \\ 0 & 0 & 1 \end{pmatrix}, \quad A(\rho) = \begin{pmatrix} 0 & 1 & 0 \\ 0 & 0 & 1 \\ 1 & 0 & 0 \end{pmatrix},$$

$$A(\lambda) = \begin{pmatrix} 0 & 1 & -1 \\ 1 & 0 & -1 \\ 0 & 0 & -1 \end{pmatrix}, \quad A(\gamma) = \begin{pmatrix} -1 & 1 & 0 \\ -1 & 0 & 1 \\ -1 & 0 & 0 \end{pmatrix},$$

$$\phi(\tau) = 1, \quad \phi(\rho) = 0, \quad \phi(\lambda) = -1, \quad \phi(\gamma) = -1.$$

6. If A were irreducible over the rationals, then any rational matrix T satisfying $TA(x) = A(x)T$ for all $x \in G$ would have to be either zero or else non-singular. However, the matrix $T = I + A(z)$ violates this conclusion. Hence A must be reducible.

7. Let $U = [\mathbf{u}_1, \ldots, \mathbf{u}_r]$. Then the vectors of $U^\perp$ are the solutions of the system of linear equations $\mathbf{w}\bar{\mathbf{u}}_i' = 0$ $(i = 1, 2, \ldots, r)$, which is of rank r. Hence dim $U^\perp = m - r$, say $U^\perp = [\mathbf{w}_1, \ldots, \mathbf{w}_{m-r}]$. We claim that the vectors $\mathbf{u}_1, \ldots, \mathbf{u}_r$, $\mathbf{w}_1, \ldots, \mathbf{w}_{m-r}$, are linearly independent; for any linear relation between them amounts to an equation of the form $\mathbf{u} + \mathbf{w} = 0$, where $\mathbf{u} \in U$, $\mathbf{w} \in U^\perp$. Multiply on the right by $\bar{\mathbf{u}}'$. Thus $\mathbf{u}\bar{\mathbf{u}}' + \mathbf{w}\bar{\mathbf{u}}' = \mathbf{u}\bar{\mathbf{u}}' = 0$, so that $\mathbf{u} = 0$, $\mathbf{w} = 0$.

Suppose now that $A(x)$ is the representation afforded by the G-module V and that

$$A(x)\bar{A}'(x) = I \quad (x \in G).$$

We assert that if U is a G-module, so also is $U^\perp$, that is, we must prove that, if $\mathbf{w} \in U^\perp$, then $\mathbf{w}A(x) \in U^\perp$. Now

$$\mathbf{w}A(x)\bar{\mathbf{u}}' = \mathbf{w}\bar{A}'(x^{-1})\bar{\mathbf{u}}' = \mathbf{w}(\overline{\mathbf{u}A(x^{-1})})' = 0,$$

because $\mathbf{u}A(x^{-1}) \in U$. Thus $U^\perp$ is a G-module complementing U, which demonstrates the complete reducibility of V.

8. Let $C = AB = (c_{ij})$. Then

$$c^{(1^2)}(i, p; j, q) = \sum_{r, s} (a_{ir}b_{rj}a_{ps}b_{sq} - a_{ir}b_{rq}a_{ps}b_{sj})$$

$$= \sum_{r, s} a_{ir}a_{ps}b(r, s; j, q)$$

$$= \sum_{r < s} a^{(1^2)}(i, p; r, s)b^{(1^2)}(r, s; j, q).$$

This amounts to the statement that $C^{(1^2)} = A^{(1^2)}B^{(1^2)}$.

We have that

$$\operatorname{tr} A^{(1^2)} = \sum_{i<j} a^{(1^2)}(i,j;i,j) = \sum_{i<j} (a_{ii}a_{jj} - a_{ij}a_{ji})$$

$$= \tfrac{1}{2} \sum_{i,j} (a_{ii}a_{jj} - a_{ij}a_{ji}) = \tfrac{1}{2}\{(\operatorname{tr} A)^2 - \operatorname{tr} A^2\}.$$

Suppose that $A(x)$ is a representation with character $\phi(x)$. Then

$$\operatorname{tr} A^{(1^2)}(x) = \tfrac{1}{2}\{(\phi(x))^2 - \phi(x^2)\}.$$

Solutions to Exercises on Chapter 2

1. We have that (p. 39)

$$\sum_{y \in G} a_{ip}(y^{-1}) a_{pj}(y) = (g/f)\delta_{ij}.$$

Multiply by $a_{js}(x)$, sum over j and use the fact that $A(y)A(x) = A(yx)$.

2. Let $K = (k_{ij})$. We are given that

$$\sum_{i,r} k_{ir} a_{ri}(x) = 0.$$

Choose p and q arbitrarily, multiply by $a_{pq}(x^{-1})$ and take inner products over G. Using equation (2.10) we deduce that $k_{pq} = 0$, that is $K = 0$.

3. (i) Since $[S_3 : H] = 3$, the representation $\sigma(x): Ht_i \to Ht_ix$ $(i = 1, 2, 3)$ certainly maps S_3 into S_3. It remains to verify that the kernel of σ is the trivial group. (See Exercise 1.1.)

(ii) Use the coset decomposition $S_3 = K \cup K(12)$, and let ϕ be the character of the representation $\tau(x): Ks_j \to Ks_jx$, where $s_1 = 1$, $s_2 = (12)$. We find that

$$\phi(1) = 2, \qquad \phi(12) = 0, \qquad \phi((123)) = 2,$$

which has the Fourier analysis

$$\phi = \chi^{(1)} + \zeta,$$

where $\chi^{(1)}$ is the trivial character.

4. Formula (2.42) can be written as $X'\bar{X} = D$, where $D = \operatorname{diag}(g/h_1, g/h_2, \ldots, g/h_k)$. This implies that X is non-singular.

If each $\chi^{(i)}$ is replaced by $\bar{\chi}^{(i)}$, the matrix X is changed into, say, PX; and if each C_α is replaced by $C_{\alpha'}$ the matrix X becomes XQ, where P and Q are permutation matrices. Thus

$$PX = (\bar{\chi}_\alpha^{(i)}), \quad XQ = (\chi_{\alpha'}^{(i)}).$$

But $\bar{\chi}_\alpha^{(i)} = \chi_{\alpha'}^{(i)}$, so that $PX = XQ$. Hence $Q = X^{-1}PX$ and therefore $\operatorname{tr} Q = \operatorname{tr} P$. The trace of a permutation matrix gives the number of fixed objects.

5. There are no elements of order 2 in G. Hence if $u \neq 1$, then $u \neq u^{-1}$. Moreover, u and u^{-1} are not conjugate. For suppose that $t^{-1}ut = u^{-1}$; it would follow that $t^{-2}ut^2 = u$, that is t^2 would belong to the centraliser $C(u)$. The order of t is necessarily odd, say $t^{2r+1} = 1$, so that $t = (t^2)^{-r}$, which would imply that $t \in C(u)$, contradicting the fact that $u \neq u^{-1}$. Thus if $u \in C_\alpha$, then $u^{-1} \in C_{\alpha'}$, where $\alpha \neq \alpha'$. Applying the character relations of the second kind to α and α' we find that

$$\sum_{i=1}^{k} (\chi_\alpha^{(i)})^2 = 0.$$

This makes it plain that not all the numbers $\chi_\alpha^{(i)}$ can be real.

6. By Exercise 2.5, the only self-inverse class is $C_1 = \{1\}$. Hence, by Exercise 2.4, there is precisely one real-valued character, namely the trivial character.

7. Let $\phi(x) = \operatorname{tr} A(x)$, $\psi(x) = \operatorname{tr} B(x)$. The hypothesis implies that $\phi(x) = \psi(x)$ for all $x \in G$. Hence, by Theorem 2.2, $A(x) \sim B(x)$.

8. Suppose that $F(x)$ is a faithful absolutely irreducible representation of the group G of order g. Let

$$Z: z_0, z_1, \ldots, z_{t-1}$$

be the centre of G. By Schur's Lemma $F(z_j) = \varepsilon_j I$, where ε_j is a gth root of unity. The set

$$E: \varepsilon_0, \varepsilon_1, \ldots, \varepsilon_{t-1}$$

consists of distinct scalars, which form a subgroup of the cyclic group of gth roots of unity. Hence E is cyclic [13, 37], and $E \cong Z$ by virtue of the map $\varepsilon_j \leftrightarrow z_j$.

9. The following equations are easily deduced from the defining relations:

$$b^{-1}ab = bab^{-1} = a^5, \quad a^{-1}ba = a^4 b,$$

whence the conjugacy classes are obtained as shown in the table.

Next, $G' = (1, a^2, a^4)$, $Z = (1, a^3)$, and $G/G' \cong C_4$, $G/Z \cong S_3$.

Lift the four linear characters of C_4 and the two-dimensional character of S_3. Observe that $\chi^{(6)} = \chi^{(2)} \chi^{(5)}$.

10. Put $D = D_{2m}$, $E = \operatorname{gp}\{a^2\}$. Then $E \lhd D$. Since $a^{-1}b^{-1}ab = a^{-2}$, we infer that $E \leq D'$. On the other hand, $|D/E| = 4$, whence D/E is Abelian and $D' \leq E$ [13, 74]. Thus $E = D'$. It follows that D has four linear characters. These are obtained by lifting the linear characters of D/D', which is isomorphic to $C_2 \times C_2$. The irreducibility of the matrix representations, for each t, can be proved by checking that $\langle \chi^{(t)}, \chi^{(t)} \rangle = 1$ or, alternatively, by applying the converse to Schur's Lemma (p. 36).

11. Put $D = D_{2m+1}$, $E = \operatorname{gp}\{a\}$. Then $E \lhd D$. Since $a^{-1}b^{-1}ab = a^{-2}$, we have that $\operatorname{gp}\{a^2\} \leq D'$. But $\operatorname{gp}\{a^2\} = \operatorname{gp}\{a\}$, because a is of odd order. Hence $E \leq D'$. On the other hand D/E is Abelian, whence $D' \leq E$. Thus $E = D'$. The elements of D/E are D, Db. The linear characters are obtained by assigning to Db the values ± 1. The matrix representations A_t, B_t are irreducible.

209

12. Put

$$P = \begin{pmatrix} 1 & -i \\ -i & 1 \end{pmatrix}.$$

13. We have to show that

$$\Theta(u_1 t^r)\Theta(u_2 t^s) = \Theta(u_1 t^r u_2 t^s) \quad (*)$$

$(u_1, u_2 \in H; r, s = 0, 1, \ldots, m-1)$. Since $H \lhd G$, we have that $t^r u_2 = u_2' t^r$, where $u_2' \in H$, and $\theta(u_2) = \theta(u_2')$ by (i). Now

$$u_1 t^r u_2 t^s = u_1 u_2' t^{r+s}.$$

Two cases arise: (a) when $r + s < m$,

$$\theta(u_1 u_2' t^{r+s}) = \theta(u_1)\theta(u_2)\varepsilon^{r+s}.$$

(b) when $r + s = m + j$, write $t^{r+s} = vt^j$. Then

$$\theta(u_1 u_2' vt^j) = \theta(u_1)\theta(u_2)\theta(v)\varepsilon^j = \theta(u_1)\theta(u_2)\varepsilon^{r+s}.$$

In both cases, this agrees with the left-hand side of (*).

14. The function $\delta(x) = \det A(x)$ is a non-trivial linear character of G. Hence, by Theorem 2.8, we have that $[G:G'] > 1$.

15. Let $[R(x)]_{p,q}$ be the (p, q)th element of $R(x)$. Then

$$[R(x)]_{p,q} = \begin{cases} 1 & \text{if } p^{-1}q = x \\ 0 & \text{if } p^{-1}q \neq x, \end{cases}$$

that is

$$[R(x)]_{p,q} = \delta_{px,q}.$$

This agrees with the formula given on p. 44, provided that we put $x = x_s$, $p = x_i$, $q = x_j$.

16. To evaluate the group determinant

$$\det\left(\sum_x R(x)\xi_x\right),$$

we may replace $R(x)$ by an equivalent representation. Use the decomposition of $R(x)$ into its irreducible constituents (p. 46).

17. This is the group determinant of the cyclic group $\text{gp}\{x\}$ order n. Its characters are $\lambda^{(r)}(x^s) = \varepsilon^{rs}$, $(r, s = 0, 1, \ldots, n-1)$.

18. Since V is generated by a and b it suffices to show that $\mathbf{u}_i a \in \{\mathbf{u}_i\}$ and $\mathbf{u}_i b \in \{\mathbf{u}_i\}$ $(i = 1, 2, 3, 4)$.

For example,

$$\mathbf{u}_1 a = a + 1 + ab + b = \mathbf{u}_1,$$
$$\mathbf{u}_3 b = b + ab - 1 - a = -\mathbf{u}_3.$$

Generally

$$\mathbf{u}_i x = \pm \mathbf{u}_i \quad (x \in V).$$

19. Let $H = \sum_{i=1}^{n} A_i \bar{A}'_i$. Then $H = \bar{H}'$, and H is a positive definite Hermitian matrix. For if $\mathbf{u}$ is an arbitrary row-vector and if we put $\mathbf{v}_i = \mathbf{u} A_i$, we obtain that

$$\mathbf{u} H \bar{\mathbf{u}}' = \sum_{i=1}^{n} |\mathbf{v}_i|^2.$$

This expression is clearly non-negative; moreover, it vanishes if and only if $\mathbf{v}_i = \mathbf{0}$ $(i = 1, 2, \ldots, n)$, whence $\mathbf{u} = 0$. By the diagonalisation theory of Linear Algebra, there exists a unitary matrix U such that

$$U^{-1} H U = D,$$

where $D = \mathrm{diag}(\alpha_1, \alpha_2, \ldots, \alpha_n)$, and each α_i is positive. Hence

$$E = \mathrm{diag}(\sqrt{\alpha_1}, \sqrt{\alpha_2}, \ldots, \sqrt{\alpha_n})$$

is a real matrix such that $E^2 = D$ and $E = E'$. Put $P = UE$. Then $H = P\bar{P}'$, because $U^{-1} = \bar{U}'$.

When H is real, U may be taken to be a real orthogonal matrix, which renders P real.

20. Since $A(x)$ is a representation, it is found that

$$A(x) H \bar{A}'(x) = \sum_y A(xy) \bar{A}'(xy) = H.$$

By Exercise 2.19, we can write $H = P\bar{P}'$. Hence the matrices $B(x) = P^{-1} A(x) P$ $(x \in G)$ satisfy

$$B(x) \bar{B}'(x) = I.$$

When $A(x)$ is real, we may take P to be real, whence $B(x)$ is real orthogonal.

21. By Exercise 2.20 every representation of a finite group (over $\mathbb{C}$) is equivalent to a unitary representation, and by Exercise 1.7, every unitary representation is completely reducible.

Solutions to Exercises on Chapter 3

1. To establish the conjugacy classes note that $a^{-1}ba = a^3 b$. Let $A = \mathrm{gp}\{a\}$. Since $a = a^{-1}b^{-1}ab$, we have that $A \leq G'$. But G/A is Abelian, whence $G' \leq A$. Thus $A = G'$. Now G/G' is generated by $G'b$, and $|G/G'| = 3$. Hence there are three linear characters, obtained by assigning to $G'b$ in turn the values $1, \omega, \omega^2$, where $\omega = \exp(2\pi i/3)$.

For the non-linear characters of G choose the linear character ξ of A, defined by $\xi(a) = \varepsilon$, where $\varepsilon = \exp(2\pi i/7)$. Put $\chi^{(4)} = \xi^G$. It is found that

$$\chi^{(4)} = (3, \varepsilon + \varepsilon^2 + \varepsilon^4, \varepsilon^3 + \varepsilon^5 + \varepsilon^6, 0, 0).$$

Let $\eta = \varepsilon + \varepsilon^2 + \varepsilon^4$. Then $\bar{\eta} = \varepsilon^3 + \varepsilon^5 + \varepsilon^6 = -1 - \eta$ and $\eta^2 + \eta + 2 = 0$. The simplicity of $\chi^{(4)}$ can be deduced either from Proposition 3.2 (p. 81) or else by verifying that $\langle \chi^{(4)}, \chi^{(4)} \rangle = 1$. To obtain $\chi^{(5)}$ replace ε by ε^3.

2. Since G is a p-group, Z is non-trivial [13, p. 56]. Also G/Z is not cyclic [13, p. 65]. Therefore G/Z is of order p^2 and hence Abelian [13, p. 66]. In fact, G/Z is generated by Za, Zb and is elementary Abelian because a^p, $b^p \in Z$. The elements of Z form conjugacy classes consisting of a single element, say $C_n = (a^{np})$ $(n = 0, 1, \ldots, p-1)$. The other conjugacy classes are $K_{\alpha,\beta} = C(a^\alpha b^\beta) = a^\alpha b^\beta Z$, where $\alpha, \beta = 0, 1, \ldots, p-1$, but $\alpha + \beta > 0$. Thus G has $p + p^2 - 1$ conjugacy classes.

The linear characters of G are obtained by lifting the p^2 linear characters of G/Z. These are constructed by assigning to aZ, bZ the values ε^r, ε^s ($r, s = 0, 1, \ldots, p-1$). Hence

$$\lambda^{(r,s)}(a^\alpha b^\beta Z) = \varepsilon^{\alpha r + \beta s}.$$

In order to find the non-linear characters, consider the normal subgroup $A = \mathrm{gp}\{a\}$ of order p^2. Use the coset decomposition

$$G = \bigcup_{\mu=0}^{p-1} Ab^\mu.$$

Let $\eta = \exp(2\pi i/p^2)$. Then $\xi(a) = \eta$ defines a linear character of A. Put $\chi = \xi^G$, thus

$$\chi(x) = \sum_{\mu=0}^{p-1} \xi(b^\mu x b^{-\mu}).$$

Note that $b^\mu a b^{-\mu} = a a^{-\mu p}$ and

$$\sum_{\mu=0}^{p-1} \eta^{-\mu p} = 0.$$

Hence $\chi(x) = 0$ if $x \notin A$, $\chi(a^\alpha) = 0$ if $(\alpha, p) = 1$, $\chi(a^{np}) = p\eta^{np} = p\varepsilon^n$ and check that $\langle \chi, \chi \rangle = 1$. Finally, replace ε by ε^t ($t = 1, 2, \ldots, p-1$), giving

$$\begin{cases} \chi^{(t)}(a^{np}) = p\varepsilon^{tn} \\ \chi^{(t)}(x) = 0 \quad \text{if } x \notin Z. \end{cases}$$

3. In this group $Z = \mathrm{gp}\{a\}$, and G/Z is elementary Abelian of order p^2 with generators bZ, cZ. The conjugacy classes are $C_\alpha = (a^\alpha)$ ($\alpha = 0, 1, \ldots, p-1$) and $K_{\beta,\gamma} = b^\beta c^\gamma Z$, where $\beta, \gamma = 0, 1, \ldots, p-1$ but $\beta + \gamma > 0$. Hence there are $p + p^2 - 1$ classes in all.

The linear characters of G are obtained by lifting the linear characters of G/Z. Assign to bZ and cZ the values ε^r, ε^s ($r, s = 0, 1, \ldots, p-1$). Hence

$$\lambda^{(r,s)}(b^\beta c^\gamma Z) = \varepsilon^{\beta r + \gamma s}.$$

For the non-linear characters consider the normal subgroup $H = \text{gp}\{a, b\}$ with the coset decomposition

$$G = \bigcup_{\mu=0}^{p-1} Hc^{\mu}.$$

Now $\xi(a^{\alpha}b^{\beta}) = \varepsilon^{\alpha}$ is a linear character of H. Put $\chi = \xi^{G}$. Then

$$\chi(x) = \sum_{\mu=0}^{p-1} \xi(c^{\mu}xc^{-\mu}).$$

Note that $c^{\mu}b^{\beta}c^{-\mu} = a^{-\beta\mu}b^{\beta}$. A simple calculation yields

$$\chi(a^{\alpha}) = p\varepsilon^{\alpha}, \chi(x) = 0 \quad \text{if } x \notin Z.$$

Finally replace ε by ε^{t} $(t = 1, 2, \ldots, p-1)$ to obtain all the non-linear characters $\chi^{(t)}$.

4. Let $|L| = l$. Then by (3.10)

$$\psi(y) = \phi^{H}(y) = (1/l) \sum_{u \in H} \phi(uyu^{-1}) \quad (y \in H),$$

where $\phi(x) = 0$ if $x \notin L$. The formula still holds for $\psi(x)$ when $x \notin H$, because each term is then equal to zero. Suppose that

$$G = \bigcup_{i=1}^{n} Ht_{i}.$$

Then

$$\psi^{G}(x) = \sum_{i=1}^{n} \psi(t_{i}xt_{i}^{-1}) = (1/l) \sum_{i=1}^{n} \sum_{u \in H} \phi(ut_{i}xt_{i}^{-1}u^{-1})$$

$$= (1/l) \sum_{w \in G} \phi(wxw^{-1}) = \phi^{G}(x).$$

5. Let $G = Ht_{1} \cup Ht_{2} \cup \ldots \cup Ht_{n}$ be a coset decomposition of G relative to H. Then

$$\{\phi_{H}\psi\}^{G}(x) = \sum_{i=1}^{n} \phi(t_{i}xt_{i}^{-1})\psi(t_{i}xt_{i}^{-1}) = \phi(x) \sum_{i=1}^{n} \psi(t_{i}xt_{i}^{-1}) = \phi(x)\psi^{G}(x).$$

6. Let $\psi^{(j)}$ $(j = 1, 2, \ldots, a)$ be the complete set of linear characters of A. Choose any simple character χ of G. Let

$$\chi_{A} = \sum_{j} a_{j}\psi^{(j)}$$

be the Fourier analysis of χ_{A} in A. At least one of the coefficients on the right is non-zero, say $a_{s} \neq 0$. Then

$$a_{s} = \langle \chi_{A}, \psi^{(s)} \rangle_{A} = \langle \chi, \psi^{(s)G} \rangle \geq 1.$$

Hence

$$\psi^{(s)G} = \ldots + a_{s}\chi + \ldots$$

is the Fourier analysis of $\psi^{(s)G}$ in G. It follows that

$$\deg \psi^{(s)G} \geq \deg \chi.$$

On the other hand, from first principles,

$$\deg \psi^{(s)G} = [G:A],$$

because $\psi^{(s)}$ is linear.

7. As on p. 83 we find that

$$C((x, y))C((x', y')) = (A(x) \otimes B(y))(A(x') \otimes B(y'))$$
$$= A(xx') \otimes B(yy') = C((xx', yy')).$$

Also

$$C((1_G, 1_H)) = I_{mn},$$

where $m = \deg A$, $n = \deg B$.

Now suppose that A and B are absolutely irreducible, with characters ϕ and ψ respectively. Thus

$$\langle \phi, \phi \rangle_G = \langle \psi, \psi \rangle_H = 1.$$

Then C has character $\phi\psi$, and it follows that

$$\langle \phi\psi, \phi\psi \rangle_w = |G|^{-1}|H|^{-1} \sum_{x,y} \phi(x)\psi(y)\bar\phi(x)\bar\psi(y)$$

$$= \left(|G|^{-1} \sum_x \phi(x)\bar\phi(x)\right)\left(|H|^{-1} \sum_y \psi(y)\bar\psi(y)\right) = 1,$$

as required.

8. We have that

$$e_r = g^{-1} \sum_{x \in G} \bar\chi(x)\{\phi(x)\}^r = g^{-1} \sum_{s=1}^{t} \sum_{x \in U_s} \bar\chi(x)\omega_s^r,$$

$$e_r = \sum_{s=1}^{t} \omega_s^r z_s. \quad (r = 0, 1, \ldots, t-1)$$

The Vandermonde determinant (p. 116)

$$\det(\omega_s^r) = \pm \prod_{r<s} (\omega_r - \omega_s)$$

is non-zero. Since $A(x)$ is faithful, $U_1 = \{1\}$ and therefore $z_1 = g^{-1}\chi(1) \neq 0$. Hence the column vector $(z_1, z_2, \ldots, z_t)'$ is non-zero and it follows that $(e_0, e_1, \ldots, e_{t-1})'$ cannot be the zero vector, because (ω_s^r) is a non-singular matrix.

9. (i) We regard $\mathbf{v} \otimes \mathbf{w}$ as a row vector of dimension mn whose components are enumerated by pairs of integers (i, j). Now $\mathbf{e}_i \otimes \mathbf{f}_j$ has a unit in the (i, j)th place and zeros elsewhere. Hence

$$\sum_{i,j} \lambda_{ij}(\mathbf{e}_i \otimes \mathbf{f}_j) = (\lambda_{11}, \lambda_{12}, \ldots, \lambda_{21}, \lambda_{22}, \ldots, \lambda_{mn}),$$

214

and this vector is zero if and only if all λ_{ij} are zero.

(ii) Let $\mathbf{v}_r = \sum_i p_{ri} \mathbf{e}_i$ and $\mathbf{w}_s = \sum_j q_{sj} \mathbf{f}_j$. Since the sets $(\mathbf{v}_r)$ and $(\mathbf{w}_s)$ are linearly independent, the matrices $P = (p_{ri})$ and $Q = (q_{sj})$ are such that $XP = 0$ implies $X = 0$ and that $YQ = 0$ implies $Y = 0$. Suppose now that

$$\sum_{r,s} h_{rs}(\mathbf{v}_r \otimes \mathbf{w}_s) = 0. \qquad (*)$$

Substituting for $\mathbf{v}_r$ and $\mathbf{w}_s$ and using the result established in (i) we see that $(*)$ is equivalent to

$$P'HQ = 0,$$

where $H = (h_{rs})$. It follows that $P'H = 0$, that is $H'P = 0$, and hence $H' = 0$. Thus $h_{rs} = 0$ for all r and s.

10. Let $Z = [\mathbf{u}_1, \ldots, \mathbf{u}_t]$ be a vector space over K and suppose that β is given by

$$\mathbf{f}_j = \sum_{k=1}^{t} b_{jk}\mathbf{u}_k.$$

Then

$$\alpha^{(2)}: V^{(2)} \to W^{(2)}, \quad \beta^{(2)}: W^{(2)} \to Z^{(2)}.$$

Hence

$$\alpha^{(2)}\beta^{(2)}: V^{(2)} \to Z^{(2)}.$$

But also

$$(\alpha\beta)^{(2)}: V^{(2)} \to Z^{(2)}.$$

Therefore $(\alpha\beta)^{(2)} = \alpha^{(2)}\beta^{(2)}$. When this equality is expressed in terms of matrices we obtain that $(AB)^{(2)} = A^{(2)}B^{(2)}$.

Solutions to Exercises on Chapter 4

1. Suppose there are n members of the party. Let $\nu(x)$ be the number of fixed objects under the permutation x, that is the number of gentlemen who pick up their own umbrellas. Then the average value of $\nu(x)$ is

$$(n!)^{-1} \sum_{x \in S_n} \nu(x) = 1,$$

by Proposition 4.2.

2. When $p_n > 0$ and $x_n = 0$, the numerator in (4.58) vanishes while the denominator remains non-zero. When $p_n = 0$ and $x_n = 0$, the last column in each determinant is $(0, 0, \ldots, 0, 1)'$. Expanding both determinants with respect to the last column and extracting the factor x_j from the jth column ($j = 1, 2, \ldots, n-1$) we find that

$$F_{n-1}^{(p)} = \det(x_j^{p_i+n-1-i})/\det(x_j^{n-1-i}).$$

$(i, j = 1, 2, \ldots, n-1)$. The process may be repeated when $p_{n-1} = p_{n-2} = \ldots = p_{m+1} = 0$, thus yielding the result.

3. Use equation (4.57) on p. 123. Suppose that $p_n = 0$. In the numerator detach

the terms that correspond to the values $j = n; i = 1, 2, \ldots, n - 1$. Observe that the last factor in the denominator reduces to unity. Thus

$$f^{(p)} = n! \prod_i (p_i + n - i) \prod_{i<j} (p_i - p_j + j - i) \left\{ \prod_i (p_i + n - i)! \right\}^{-1}$$

$$= n! \prod_{i<j} (p_i - p_j + j - i) \left\{ \prod_i (p_i + n - 1 - i)! \right\}^{-1}$$

$(i, j = 1, 2, \ldots, n - 1)$. Repeat the process if $p_{n-1} = 0$ etc.

4. Apply the result of Exercise 4.2 with $m = 2$. Thus put $x_3 = \ldots = x_n = 0$. Then $F_2^{(p)} = 0$ when p has more than three positive terms. When $p = [n - t, t]$,

$$F_2^{(p)} = (x_1^{n-t+1} x_2^t - x_1^t x_2^{n-t+1})/(x_1 - x_2).$$

Substitute in (4.58) and put $x = x_2/x_1$.

The particular formulae are obtained by comparing the coefficients of x, x^2 and x^3 respectively.

5. We have that

$$\frac{f'(t)}{f(t)} = \frac{d}{dt} \sum_i \log(1 - x_i t) = \sum_i \frac{-x_i}{1 - x_i t}$$

$$= -(s_1 + s_2 t + s_3 t^2 + \ldots).$$

Multiply by $f(t)$ and compare coefficients of t^{r-1} to obtain the rth recurrence relation. Write down the first r of these relations and regard them as a system of linear equations for $c_1, c_2, \ldots, c_r$. Apply Cramer's Rule to solve for c_r.

6. Differentiating the generating function

$$\{f(t)\}^{-1} = \sum_{r=0}^{\infty} w_r t^r$$

we find that

$$-f'(t)\{f(t)\}^{-2} = \sum_{r=1}^{\infty} r w_r t^{r-1} = -(f'(t)/f(t))\{f(t)\}^{-1} = \left(\sum_\alpha s_\alpha t^{\alpha-1} \right) \left(\sum_\beta w_\beta t^\beta \right).$$

Compare coefficients of t^{r-1} to obtain the recurrence relation. Regard the first r relations as a system of r linear equations for $w_1, w_2, \ldots, w_r$ and solve for w_r by means of determinants.

7. Put $p = [n - 2, 1^2]$. Then, by (4.70),

$$F^{(p)} = \begin{vmatrix} w_{n-2} & w_{n-1} & w_n \\ 1 & w_1 & w_2 \\ 0 & 1 & w_1 \end{vmatrix} = (w_1^2 - w_2)w_{n-2} - w_1 w_{n-1} + w_n$$

$$= \tfrac{1}{2}(s_1^2 - s_2)w_{n-2} - s_1 w_{n-1} + w_n.$$

216

Using (4.65) and writing (4.60) in the form

$$F^{(p)} = \sum_{\|\alpha\|=n} g(\alpha)\chi_\alpha^{(p)}s_\alpha$$

we find that

$$g(\alpha_1, \alpha_2, \ldots)\chi_\alpha^{(p)} = \tfrac{1}{2}g(\alpha_1 - 2, \alpha_2, \ldots)$$
$$- \tfrac{1}{2}g(\alpha_1, \alpha_2 - 1, \ldots) - g(\alpha_1 - 1, \alpha_2, \ldots)$$
$$+ g(\alpha_1, \alpha_2, \ldots).$$

Hence

$$\chi_\alpha^{(p)} = \tfrac{1}{2}\alpha_1(\alpha_1 - 1) - \alpha_2 - \alpha_1 + 1$$
$$= \tfrac{1}{2}(\alpha_1 - 1)(\alpha_1 - 2) - \alpha_2.$$

8. If the rank of p is equal to unity, then $p_1 \geq 1$ and $p_2 < 2$. Thus $p_2 \leq 1$, and therefore $p_j \leq 1$ when $j \geq 2$.

9. We use the formula (4.56), namely

$$s_\alpha \Delta = \sum_p \chi_\alpha^{(p)} V^{(p)}.$$

A typical term of $V^{(p)}$ can be written as

$$\pm(x\sigma|p + e),$$

where $\sigma \in S_n$ and $e: n - 1, n - 2, \ldots, 1, 0$. Similarly, each term of Δ is of the form

$$\pm(x\rho|e) \quad (\rho \in S_n).$$

Suppose that the elements of C_α are products of cycles of degrees $r_1, r_2, \ldots r_t$. Then $s_\alpha = s_{r_1}s_{r_2}\ldots s_{r_t}$ and s_α is a sum of terms $(x\lambda|a)$, where $\lambda \in S_n$ and a is a partition of n which contains at most t positive terms. Hence $s_\alpha \Delta$ is a sum of terms

$$\pm(x\lambda|a)(x\rho|e) = \pm(x\pi|q + e), \tag{*}$$

say. Retaining products with distinct exponents only we may assume that $q + e$ is a s.d. partition of $\tfrac{1}{2}n(n + 1)$, whence $q : q_1 \geq q_2 \geq \ldots \geq q_n$ is a partition of n. It follows from (*) that the inequality

$$q_i + e_i \geq n, \text{ that is, } q_i \geq i$$

cannot hold unless $i \leq t$, and the rank of q does not exceed t. Thus $\chi_\alpha^{(p)} = 0$ if the rank of p is greater than t.

10. Assign to $x_1, x_2, \ldots, x_n$ those values which imply that $s_1 = s_2 = \ldots = s_{n-1} = 0$, $s_n = 1$, that is let them be the roots of

$$n\xi^n - 1 = 0.$$

Then $w_0 = 1$, $w_2 = \ldots = w_{n-1} = 0$, $w_n = 1/n$.

Let p be a partition that consists of u non-zero terms, and suppose that $F^{(p)}$ reduces to $F_*^{(p)}$ when the above values are substituted. Then by (4.60)

$$F_*^{(p)} = \frac{1}{n}\chi_\gamma^{(p)}, \tag{1}$$

217

because $h_\gamma = (n-1)!$. But the first row of

$$\det(w_{p_i-i+j})_* \quad (i, j = 1, 2, \ldots, u) \tag{2}$$

reduces to zero unless

$$p_1 - 1 + u \geqslant n.$$

Since $p_j \geqslant 1$ $(j = 2, 3, \ldots, u)$ we have that

$$n = p_1 + p_2 + \ldots + p_u \geqslant (n - u + 1) + (u - 1) = n.$$

A contradiction arises unless the equality holds in all cases, so that $p_1 = n - u + 1$, $p_j = 1$ $(j \geqslant 2)$. Putting $u - 1 = t$ we see that $p = [n - t, 1^t]$ is a partition of rank unity. The determinant (2) now becomes

$$\begin{vmatrix} 0 & 0 & \ldots & 0 & w_n \\ 1 & 0 & \ldots & 0 & 0 \\ 0 & 1 & \ldots & 0 & 0 \\ & & & & \\ 0 & 0 & \ldots & 1 & 0 \end{vmatrix} = (-1)^{u+1} w_n = (-1)^t/n.$$

The result follows by equating (1) and (2).

11. There are 11 partitions of 6. The formulae given in Exercises 4.4 and 4.7 suffice to obtain the characters which correspond to the partitions

$$[6], [51], [42], [3^2], [4\,1^2]$$

and their conjugates

$$[1^6], [2\,1^4], [2^2 1^2], [2^3], [3\,1^3].$$

It only remains to find $\chi^{[3\,2\,1]}$, which is self-conjugate. Its degree is equal to 16, as can be inferred from Exercise 4.3 or (2.26); the values for the other classes are readily obtained from the column orthogonality of the character table.

Solutions to Exercises on Chapter 5

1. We use the formula

$$\sum_{i=1}^{k} (f^{(i)})^2 = pq,$$

where $f^{(i)}$ is the degree of the ith irreducible representation. Without loss of generality assume that $f^{(1)} \leqslant f^{(2)} \leqslant \ldots \leqslant f^{(k)}$. Since $|G/G'| = p$, there are precisely p linear characters. By Theorem 4.4, we have that $f^{(i)} | pq$. When $f^{(i)} > 1$, this implies that $f^{(i)} = p$, because it is obvious that $(f^{(i)})^2 < pq$. Thus

$$f^{(1)} = f^{(2)} = \ldots = f^{(p)} = 1, \quad f^{(p+1)} = \ldots = f^{(k)} = p.$$

Substituting in the above formula we find that

$$p + (k - p)p^2 = pq,$$

whence $k = p + (q - 1)p^{-1}$.

2. For a fixed value of j let $F^{(j)}$ be the irreducible representation with character $\chi^{(j)}$. Extend $F^{(j)}$ to G_C. Then (p. 147) $F^{(j)}(c_\alpha) = (h_\alpha \chi_\alpha^{(j)} / f^{(j)}) I_f$. Operate on the structure equations with $F^{(j)}$ omitting the matrix I_f. Hence

$$h_\alpha h_\beta \chi_\alpha^{(j)} \chi_\beta^{j} / f^{(j)} = \sum_{\sigma=1}^{k} a_{\alpha\beta\sigma} h_\sigma \chi_\sigma^{(j)}.$$

Multiply by $\bar{\chi}_\gamma^{(j)}$ and sum over j. The result follows by virtue of the character relations of the second kind (p. 51).

Next,

$$e^{(i)} e^{(j)} = (f^{(i)} f^{(j)} / g^2) \sum_{\alpha,\beta,\gamma} \bar{\chi}_\alpha^{(i)} \bar{\chi}_\beta^{(j)} a_{\alpha\beta\gamma} c_\gamma.$$

Substitute the expression for $a_{\alpha\beta\gamma}$ found above and use the character relations of the first kind (p. 39). Then

$$e^{(i)} e^{(j)} = \delta_{ij} \sum_\gamma \bar{\chi}_\gamma^{(i)} c_\gamma = \delta_{ij} e^{(i)}.$$

3. Using Theorem 4.4 and the formula (2.26) we infer that the degree f of an absolutely irreducible representation satisfies $f | p^m$ and $f^2 < p^m$, where $p^m = |G|$. Thus, when $m - 4$, it follows that $f = 1$ or $f = p$.

4. By Exercise 2.6 the simple characters may be listed as

$$\chi^{(1)}, \chi^{(2)}, \bar{\chi}^{(2)}, \ldots, \chi^{(l)}, \bar{\chi}^{(l)},$$

where $k = 2l - 1$. Evidently $\chi^{(i)}$ and $\bar{\chi}^{(i)}$ have the same degree $f^{(i)}$. Hence the formula (2.26) becomes

$$g = 1 + 2 \sum_{i=2}^{l} (f^{(i)})^2.$$

Now $f^{(i)}$ is odd by virtue of Theorem 4.4, say

$$f^{(i)} = 4n_i \pm 1 \quad (i = 2, \ldots, l).$$

Thus

$$g = 1 + 2 \sum_{i=2}^{l} (16n_i^2 \pm 8n_i + 1),$$

$$g = 1 + 2(l - 1) + 16N,$$

where N is an integer. Therefore $g - k = 16N$.

Solutions to Exercises on Chapter 6

1. The $2n$ elements of D_n are $a^k, a^k b$ $(k=0, 1, \ldots, n-1)$. We have that $b^{-1}a^k b = (b^{-1}ab)^k = a^{-k}$; thus $a^k \sim a^{-k}$. Also $ba^k = a^{-k}b$, whence $(a^k b)^2 = a^k b a^k b = a^k a^{-k} b^2 = 1$. Therefore $a^k b = (a^k b)^{-1}$. Hence $x \sim x^{-1}$ for all x.

2. It suffices to prove that $b \not\sim b^{-1}$. Each conjugate of b is of the form

$$a^k b a^{-k} = a^{2k} b,$$

because $aba^{-1} = a^2 b$. Suppose that, for some k,

$$a^{2k} b = b^{-1} = b^3;$$

then

$$a^{2k} = b^2 = a^n.$$

Hence

$$2k \equiv n \pmod{2n},$$

which implies that n is even, contrary to hypothesis.

3. There are $(n-1)!$ distinct cycles of order n in S_n; they form the conjugacy class of z. Hence the centraliser of z is of order $n!/(n-1)! = n$, and therefore contains no elements besides z^k $(k=0, 1, \ldots, n-1)$.

In S_n one solution of

$$t^{-1}zt = z^{-1} \qquad (*)$$

is

$$t = \begin{pmatrix} 1 & 2 & \ldots & n \\ n & n-1 & \ldots & 1 \end{pmatrix}.$$

This permutation involves

$$(n-1)+(n-2)+\ldots+2+1 = \tfrac{1}{2}n(n-1)$$

inversions [13, p. 134]. When $n = 4m-1$, this number is odd; so t is an odd permutation. The most general solution of $(*)$ is

$$(z^k t)^{-1} z (t z^k) = z^{-1}.$$

But tz^k, too, is odd. It follows that z and z^{-1} are not conjugate in A_n.

4. We write $f^{(i)} = \chi^{(i)}(1)$. By Theorem 2.3, $f^{(i)} \leqslant 4$ because $g = 24$. We consider the columns in turn.

(i) $\alpha = (123)$; $p = 3$.

$$\sum_{i=1}^{5} (\chi_\alpha^{(i)})^2 = \tfrac{24}{8} = 3.$$

So

$$\chi_\alpha^{(i)} = 1, 1, \pm 1, 0, 0$$

in some order. There must be at least one negative entry because

$$\sum_i f^{(i)} \chi_\alpha^{(i)} = 0.$$

Hence

$$\chi_\alpha^{(3)} = -1, \quad \chi_\alpha^{(4)} = \chi_\alpha^{(4)} = 0.$$

By Proposition 6.3,

$$f^{(3)} = 2.$$

Next, we have that

$$(f^{(4)})^2 + (f^{(5)})^2 = 24 - 1 - 1 - 5 = 18,$$

whence

$$f^{(4)} = f^{(5)} = 3.$$

The columns headed (1) and (123) are now complete; but, as yet, there is nothing to distinguish between $\chi^{(4)}$ and $\chi^{(5)}$.

(ii) $\beta = (1234); p = 2$. Since

$$(\chi_\beta^{(3)})^2 + (\chi_\beta^{(4)})^2 + (\chi_\beta^{(5)})^2 = 2,$$

one of the terms on the left is equal to zero and the other two are equal to unity. By Proposition 6.3, we must have that

$$\chi_\beta^{(3)} = 0, \quad \chi_\beta^{(4)} = \pm 1, \quad \chi_\beta^{(5)} = \pm 1.$$

Orthogonality with the first column requires that the last two values are of opposite sign. Accordingly, we fix the notation in such a way that

$$\chi_\beta^{(4)} = -1, \quad \chi_\beta^{(5)} = 1.$$

(iii) $\gamma = (12); p = 2$. The argument is similar to (ii), except that the last two entries must have signs opposite to those for β in order to satisfy the orthogonality relations with the columns of (1) and (β).

(iv) $\delta = (12)(34); p = 2$. The entries in this last remaining column can be found by applying the orthogonality relations either for the rows or for the columns. Thus

$$\chi_\delta^{(3)} = 2, \quad \chi_\delta^{(4)} = \chi_\delta^{(5)} = -1.$$

Solutions to Exercises on Chapter 7

1. When A is positive definite, all its latent roots are positive. Hence there are positive real numbers l_i such that $l_i^2 = \lambda_i$ $(l = 1, 2, \ldots, n)$. Let $L = \mathrm{diag}(l_1, l_2, \ldots, l_n) = L'$. By the Principal Axes Theorem,

$$R'AR = L^2 = LL',$$

whence

$$A = RLL'R' = PP',$$

where

$$P = RL.$$

2. It is easy to verify that A and B satisfy the defining relations of Δ_n, namely

$$A^{2n} = I, \quad A^n = (AB)^2 = B^2.$$

Hence A and B generate a representation.

In order to prove that the representation is irreducible over $\mathbb{C}$, it suffices to show that the only matrices that commute with both A and B, are the scalar multiples of the unit matrix (Theorem 1.5). Alternatively, one could check that $\langle \chi, \chi \rangle = 1$.

It is readily shown that

$$ABA' = B, \quad BBB' = B.$$

Hence B is a skew-symmetric invariant. The representation is therefore of type II.

3. The representation has a non-singular skew-symmetric invariant. All skew-symmetric matrices of odd degree are singular.

4. The matrix

$$\sum_y F(y^2)$$

commutes with each matrix $F(x)$ $(x \in G)$, because

$$x^{-1}y^2x = (x^{-1}yx)^2.$$

Hence

$$\sum_y F(y^2) = \lambda I, \tag{1}$$

where λ is a scalar. On taking the trace of each side and using (7.37) we find that

$$cg = \lambda f. \tag{2}$$

After multiplying (1) on the left by $F(x)$ and taking traces we obtain that

$$\chi(xy^2) = \lambda\chi(x),$$

where λ is as given in (2).

5. For each irreducible character $\chi^{(i)}$,

$$\sum_{y \in G} \chi^{(i)}(y^2) = c_i g. \tag{3}$$

Suppose that, as y ranges over G, we have that $y^2 = u$ precisely $\zeta(u)$ times. Then we can write (3) as

$$\sum_{u \in G} \zeta(u)\chi^{(i)}(u) = c_i g. \tag{4}$$

We note that $\zeta(u)$ is a class function; for if $v = t^{-1}ut$, then the equations $u = y^2$ and $v = (t^{-1}yt)^2 = z^2$ have the same number of solutions in y and z respectively.

We shall write $\zeta(u) = \zeta_\alpha$ when u belongs to the class α. Hence (4) becomes

$$\sum_\alpha h_\alpha \zeta_\alpha \chi_\alpha^{(i)} = c_i g,$$

where h_α is the number of elements in α.

Next we use the orthogonality relations of the second kind (2.42) to solve for ζ. Thus

$$\zeta_\beta = \sum_i c_i \bar{\chi}_\beta^{(i)}.$$

Since $c_i = 0$ when $\chi^{(i)}$ is complex, we may replace $\bar{\chi}^{(i)}$ by $\chi^{(i)}$ in the sum.

When $\beta = (1)$, the result reduces to

$$\zeta_1 = \sum_i c_i f^{(i)}.$$

Since $\zeta_1 \geqslant 1$, we have that

$$\sum_I f > \sum_{II} f,$$

where the sums refer to representations of types I and II respectively.

6. When $n = 1$, the formula (7.43) reduces to (7.37). Assuming that $n \geqslant 2$, apply (7.42) with $x = y_1^2 y_2^2 \ldots y_{n-1}^2$ and $y = y_n^2$. On summing over $y_1, y_2, \ldots, y_n$ we obtain the recurrence relation

$$\sum_{y_1, \ldots, y_n} \chi(y_1^2 \ldots y_{n-1}^2 y_n^2) = \frac{cg}{f} \sum_{y_1, \ldots, y_{n-1}} \chi(y_1^2 \ldots y_{n-1}^2),$$

whence (7.43) follows by induction on n.

For a given element u of G, let $\zeta(n; u)$ denote the number of solutions of the equation

$$y_1^2 y_2^2 \ldots y_n^2 = u. \tag{5}$$

Evidently $\zeta(n; u)$ is a class function on G, because (5) holds if and only if

$$(t^{-1} y_1 t)^2 (t^{-1} y_2 t)^2 \ldots (t^{-1} y_n t)^2 = t^{-1} u t.$$

We write

$$\zeta(n; u) = \zeta_\alpha(n)$$

when u belongs to the class α. Hence for each irreducible character $\chi^{(i)}$ we have that

$$\sum_\alpha h_\alpha \zeta_\alpha(n) \chi_\alpha^{(i)} = c_i^n g^n (f^{(i)})^{1-n}.$$

On multiplying throughout by $f^{(i)}$ and summing over i we obtain that

$$\sum_\alpha \left\{ h_\alpha \zeta_\alpha(n) \sum_i f^{(i)} \chi_\alpha^{(i)} \right\} = g^n \sum_i c_i^n (f^{(i)})^{2-n}.$$

Now (7.44) follows because by the orthogonality relations (2.42) we have that

$$\sum_i f^{(i)} \chi_\alpha^{(i)} = g \quad \text{or} \quad 0,$$

according as $\alpha = (1)$ or $\alpha \neq (1)$.

The equations

$$y_1^2 y_2^2 = 1 \quad \text{and} \quad x^2 = y^2$$

have the same number of solutions, namely

$$\zeta_1(2) = g \sum_i c_i^2.$$

Evidently this number is equal to g times the total number of irreducible real characters of types I or II.

BIBLIOGRAPHY

[1] H. Bender, A group theoretic proof of Burnside's $p^a q^b$-theorem. *Math. Z.* **126** (1972), 327–38.

[2] M. Burrow, *Representation Theory of Finite Groups*. Academic Press, 1965.

[3] R. D. Carmichael, *Groups of Finite Order*. Dover, 1956.

[4] A. H. Clifford, Representations induced in an invariant subgroup. *Annals Math.* **38** (1937), 533–50.

[5] C. W. Curtis and I. Reiner, *Representation Theory of Finite Groups and Associative Algebras*. John Wiley and Sons, 1962.

[6] J. D. Dixon, *Problems in Group Theory*. Blaisdell, 1967.

[7] L. Dornhoff, *Group Representation Theory*, parts A and B. Marcel Dekker, 1971.

[8] W. Feit, *Characters of Finite Groups*. W. A. Benjamin, 1967.

[9] G. Frobenius, Über die Charaktere der symmetrischen Gruppe. *Sitz. Ber. Preuss. Akad. Berlin* (1900), 516–34.

[9a] G. Frobenius, Über die charakteristischen Einheiten der symmetrischen Gruppe. *Sitz. Ber. Preuss. Akad. Berlin* (1903), 328–58.

[9b] G. Frobenius and I. Schur, Über die reellen Darstellungen der endlichen Gruppen. *Sitz. Ber. Preuss. Akad. Berlin* (1906), 186–208.

[10] Marshall Hall, Jr, *The Theory of Groups*. Macmillan, 1959, Chapter 16.

[11] G. H. Hardy and E. M. Wright, *Introduction to the Theory of Numbers*. Oxford, 1954.

[12] A. Kerber, *Representations of Permutation Groups I*. Springer, 1971.

[13] W. Ledermann, *Introduction to Group Theory*. Longman, 1976.

[14] D. E. Littlewood, *The Theory of Group Characters and Matrix Representations of Groups*. Oxford, 1940.

[15] G. W. Mackey, On induced representations of groups. *Am. J. Math.* **73** (1951), 576–92.

[16] S. MacLane and G. Birkhoff, *Algebra*. Macmillan, 1967.

[17] Morris Newman, *Matrix Representations of Groups*. National Bureau of Standards, Applied Mathematics Series 60, 1968.

[18] H. Pollard, *The Theory of Algebraic Numbers*. Carus Mathematical Monographs 9, Mathematical Association of America, 1950.

[19] G. de B. Robinson, *Representation Theory of the Symmetric Group*. Edinburgh University Press, 1961.

[20] D. E. Rutherford, *Substitutional Analysis*. Edinburgh University Press, 1948.

[21] I. Schur, Neue Begründung der Theorie der Gruppencharaktere. *Sitz. Ber. Preuss. Akad. Berlin* (1905), 406–32.

[21a] I. Schur, *Die algebraischen Grundlagen der Darstellungstheorie der Gruppen*. Graph. Anstalt Gebr. Frey und Katz, Zürich, 1936.

[22] J.-P. Serre, *Représentations linéaires des groupes finis*. Hermann, Paris, 1967.

[22a] J.-P. Serre, *Linear Representations of Finite Groups*. Springer, 1971.

[23] I. Stewart, *Galois Theory*. Chapman and Hall, 1973.

INDEX

absolute irreducibility, 37
adapted basis, 11
algebraic conjugates, 142
algebraic integer, 143
algebraic number, 141
alternant, 126, 188
alternating polynomial, 112
anti-symmetric matrix, 92
anti-symmetric polynomial, 112
anti-symmetric tensor, 92
automorphism, 7
automorphism representation, 7

bilinear invariant, 177
Brioschi's formulae, 139
Burnside's (p, q)-theorem, 148

Cauchy's class formula, 108
Cauchy's determinant, 118, 187
Cayley's theorem, 1
centraliser, 71
character
  alternating, 4
  compound, 40
  deduced, 73
  generalised, 110
  induced, 70
  linear, 62
  natural, 18
  restricted, 73
  simple, 40
character relations of the first kind, 39
character relations of the second kind, 51
class function, 10
Clifford's Theorem, 82
commutant algebra, 27
complete symmetric functions, 125
composition series, 16
compound matrix (second), 36, 93
conjugacy class, 10
cyclotomic polynomial, 162

degree, 3, 120
difference product, 116
dimension, 3
direct product, 29
direct sum, 23, 29

eigenvalue (latent root), 50
endomorphism algebra, 33
Euler function, 161

Fourier analysis, 41
Frobenius group, 152

G-endomorphism, 33
G-homomorphism, 24
G-module, 8, 10, 23
generating function, 115
group
  alternating, 61, 75
  dicyclic, 65, 184
  dihedral, 61, 65, 66
  doubly transitive, 103
  dual, 55
  Frobenius, 152
  inertia, 80
  quaternion, 62
  transitive, 101
group algebra, 42
group determinant, 67
group matrix, 67
ground field, 3

induced matrix (second), 100
inner product, 38
invariant
  Hermitian, 174
  bilinear, 177
inverse classes, 52

Jacobi, C. G. J., 126, 134, 191

kernel, 58
Kronecker product, 29, 99
Kronecker power, 99

latent root (eigenvalue), 50
lifting process, 58

Mackey's Theorem, 94
Maschke's Theorem, 21

Newton's formulae, 139

orthogonality relations
  first kind, 39
  second kind, 51

p-component, 198
p-regular component, 198
partition, 107
  conjugate, 129
  graph, 128
  rank of, 140
  self-conjugate, 129
  strictly decreasing (s.d.), 112
polynomial
  alternating, 112
  minimal, 141
  monic, 141
  skew-symmetric, 112
  symmetric, 111
principle of linearity, 5

rank of partition, 140
Reciprocity Theorem, 74
reducibility, 12
  complete, 23
representation, 1
  automorphism, 4
  completely reducible, 23
  contragredient, 16
  deduced, 73
  equivalent, 4
  faithful, 4
  induced, 70
  irreducible, 12, 42

representation – *cont.*
  linear, 4
  matrix, 3
  natural, 18
  orthogonal, 173
  permutation, 17
  principal, 1
  reducible, 12
  right-regular, 2, 44
  trivial, 4
  types, I, II, III, 171

S-function, 124
Schläflian matrix (second), 90
Schur function, 124
Schur's Lemma, 24, 26
self-inverse class, 52
skew-symmetric polynomial, 112
specification, 107
stabiliser, 102

tensor
  anti-symmetric, 92
  rank 2, 88
  skew-symmetric, 92
  symmetric, 89
tensor product, 29, 87
trace, 9
transversal, 69

Vandermonde determinant, 116, 186

Wronski functions, 125

227